The Psychology of Problem Solving

Problem Solving

The Background to Successful
Mathematics Thinking

Problem Solving in Mathematics and Beyond

Print ISSN: 2591-7234
Online ISSN: 2591-7242

Series Editor: Dr. Alfred S. Posamentier
Distinguished Lecturer
New York City College of Technology - City University of New York

There are countless applications that would be considered problem solving in mathematics and beyond. One could even argue that most of mathematics in one way or another involves solving problems. However, this series is intended to be of interest to the general audience with the sole purpose of demonstrating the power and beauty of mathematics through clever problem-solving experiences.

Each of the books will be aimed at the general audience, which implies that the writing level will be such that it will not engulfed in technical language — rather the language will be simple everyday language so that the focus can remain on the content and not be distracted by unnecessarily sophiscated language. Again, the primary purpose of this series is to approach the topic of mathematics problem-solving in a most appealing and attractive way in order to win more of the general public to appreciate his most important subject rather than to fear it. At the same time we expect that professionals in the scientific community will also find these books attractive, as they will provide many entertaining surprises for the unsuspecting reader.

Published

For the complete list of volumes in this series, please visit www.worldscientific.com/series/psmb

The Psychology of Problem Solving
The Background to Successful Mathematics Thinking

Alfred S. Posamentier
City University of New York, USA

Gary Kose
Long Island University, USA

Danielle Sauro Virgadamo
Kennedy Krieger Institute, USA

Kathleen Keefe-Cooperman
Long Island University, USA

World Scientific

NEW JERSEY · LONDON · SINGAPORE · BEIJING · SHANGHAI · HONG KONG · TAIPEI · CHENNAI · TOKYO

Published by

World Scientific Publishing Co. Pte. Ltd.

5 Toh Tuck Link, Singapore 596224

USA office: 27 Warren Street, Suite 401-402, Hackensack, NJ 07601

UK office: 57 Shelton Street, Covent Garden, London WC2H 9HE

Library of Congress Cataloging-in-Publication Data

Names: Posamentier, Alfred S., author.

Title: The psychology of problem solving : the background to successful mathematics thinking /
by Alfred S. Posamentier (City University of New York, USA) [and three others].

Description: New Jersey : World Scientific, 2019. | Series: Problem solving in
mathematics and beyond ; volume 12 | Includes index.

Identifiers: LCCN 2019020071 | ISBN 9789811205705 (hc : alk. paper)

Subjects: LCSH: Problem solving--Methodology. | Mathematics

Classification: LCC QA63 .P795 2019 | DDC 153.4/3--dc23

LC record available at https://lccn.loc.gov/2019020071

British Library Cataloguing-in-Publication Data

A catalogue record for this book is available from the British Library.

For any available supplementary material, please visit
https://www.worldscientific.com/worldscibooks/10.1142/11426#t=suppl

Desk Editors: V. Vishnu Mohan/Tan Rok Ting

Typeset by Stallion Press
Email: enquiries@stallionpress.com

Printed in Singapore

Preface

We live in a continually evolving digital world where people can approach problem solving in a variety of ways and use the many digital tools now available to us. Whether it is something simplistic such as determining the tip for a server at a restaurant or the more involved task of deciding where to go on vacation, we can easily use digital tools (e.g., a tip calculator or travel website) to find answers. Digital and electronic tools provide us with convenient ways to solve problems, but these problems can be seen as a few examples of how we often avoid problem solving. This avoidance is often seen in the area of mathematics. The avoidance of problem solving begins early for many who are faced with difficult math problems in school and then expands to a habitual pattern of avoidance through life. This pattern of avoidance makes the very task of problem solving appear more daunting than it actually is.

The authors of this book came together to address the psychology of problem solving. Dr. Alfred Posamentier is a mathematics educator whose love for mathematics is the motivating force behind his desire to help others develop a genuine appreciation for mathematics and strengthen problem-solving skills there and beyond. Dr. Gary Kose, Dr. Danielle Sauro Virgadamo, and Dr. Kathleen Keefe-Cooperman are psychologists who understand how people cognitively and behaviorally develop difficulties and anxiety regarding problem solving. Ineffective patterns of thinking start in youth and become fixed over time. Dr. Kose and Dr. Virgadamo provide the historical background of problem solving as well as the psychological factors that contribute to our cognitive difficulties. These patterns of thought become our "go-to" way of thinking when a math exercise is presented or when faced with any problem scenario that causes us to cognitively or emotionally "shut down" or become overwhelmed. Both the cognitive strategies necessary to succeed as well as practice exercises are provided throughout the book to improve your skills. Answers are also provided so that you are able to analyze which problems were performed correctly and where difficulties arose. The exercises range from simple to quite difficult, but it is the willingness to learn from mistakes that will make you a better problem solver.

The book is arranged to help the reader understand the history of problem solving. It is hard to imagine that a person suffered through math problems centuries ago in much the same way as a child today would approach an algebra test. Yet, there is a long history of literature that addresses the difficulty of solving problems. Most of these early volumes focused on reaching accurate solutions and not until modern times was there a shift toward understanding the psychological process involved when solving a problem. A brief view of the history of problem solving will provide a base from which to grow in learning new ways to think about problem solving.

Much of the book focuses on strategies to improve your ability to unravel these complexities with clear steps provided. The authors move from the historical perspective in Chapter 1 to strategies for problem solving in Chapter 2 and understanding the factors that keep you from finding a solution. Many people do not often stop and think about strategical tactics before jumping in to solve a problem, so these chapters provide the reader with these skills that can be applied and practiced through the exercises provided.

Chapter 3 further explores the concepts of algorithms and heuristics with the goal that you will understand how to use these practical methods for learning. Distortions in thinking when approaching problem solving are a major barrier to success. Understanding the types of cognitive falsehoods that we apply in daily life is key to improving our skills. Chapter 4 addresses the cognitive and emotional barriers we often accidentally integrate into our daily lives, which prevent us from succeeding. Anxiety is a word that is thrown about daily without a true realization of how seriously it impacts our willingness to leave our comfort zone and tackle something new. The same is true for feelings of disinterest, which often occur when we do not understand the importance or utility of a particular skill. This chapter will provide the reader with the tools needed to overcome these feelings, so as to be more willing to face possible failure. We hope to help you accept the notion of being willing to fail in order to succeed.

We then transition to understanding the areas of focused attention and working memory, in order to help the reader to improve his or her ability to solve problems more accurately and efficiently. These strategies are presented in Chapter 5 along with more practice exercises. Being able to learn about the factors that impede our success and then tackle those problems through practice is one of the keys to becoming a better lifelong problem solver. The final chapter of the book focuses on the concepts of intuitive and deliberate thinking. Gaining an understanding about these two ways of thinking will improve your ability to lessen errors in judgment and to tackle problems that you find in everyday life.

A fear of math and the inability to effectively problem solve are not fixed by just one method. Humans are complex, and how we think about everyday life is molded by a combination of thoughts and feelings that develop over time. This book provides the reader with an understanding of the many cognitive processes that negatively affect our ability to problem solve and the new ways of thinking that can be practiced to overcome these ineffective patterns. You will also have many opportunities to practice your skills with the exercises presented in each chapter. We truly hope that this book will provide you with the understanding and tools to make you a better problem solver.

About the Authors

 **Alfred S. Posamentier** is currently Distinguished Lecturer at New York City College of Technology of the City University of New York. Previously, he was the Executive Director for Internationalization and Sponsored Programs at Long Island University, New York. This was preceded by a five-year period where he was Dean of the School of Education and tenured Professor of Mathematics Education at Mercy College, New York. He is now also Professor Emeritus of Mathematics Education at The City College of the City University of New York, and former Dean of the School of Education, where he was tenured for 40 years. He is the author and co-author of more than 60 mathematics books for teachers, secondary and elementary school students, and the general readership. Dr. Posamentier is also a frequent commentator in newspapers and journals on topics relating to education. After completing his B.A. degree in mathematics at Hunter College of the City University of New York in 1964, he took a position as a teacher of mathematics at Theodore Roosevelt High School (Bronx, NY), where he focused his attention on improving the students' problem-solving skills and at the same time enriching their instruction far beyond what the traditional textbooks offered. During his six-year tenure there, he also developed

the school's first mathematics teams (both at the junior and senior level). He is still involved in working with mathematics teachers and supervisors, nationally and internationally, to help them maximize their effectiveness. During this time, he earned an M.A. degree at the City College of the City University of New York in 1966.

Immediately upon joining the faculty of the City College in 1970, he began to develop in-service courses for secondary school mathematics teachers, including such special areas as recreational mathematics and problem solving in mathematics. As Dean of the City College School of Education for 10 years, his scope of interest in educational issues covered the full gamut. During his tenure as Dean, he took the School of Education from the bottom of the New York State rankings to the top with a perfect NCATE accreditation assessment in 2009. He achieved the same success in 2014 at Mercy College, which received both NCATE and CAEP accreditation during his leadership as Dean of the School of Education.

In 1973, Dr. Posamentier received his Ph.D. from Fordham University (New York) in mathematics education and has since extended his reputation in mathematics education to Europe. He has been visiting professor at several European universities in Austria, England, Germany, Czech Republic, and Poland, while at the University of Vienna he was Fulbright Professor (1990). In 1989, he was awarded an *Honorary Fellow* at the South Bank University (London, England). In recognition of his outstanding teaching, the City College Alumni Association named him *Educator of the Year* in 1994, and in 2009. New York City had the *day*, May 1, 1994, named in his honor by the President of the New York City Council. In 1994, he was also awarded the Grand Medal of Honor from the Republic of Austria, and in 1999, upon approval of Parliament, the President of the Republic of Austria awarded him the title of *University Professor of Austria*. In 2003, he was awarded the title of *Ehrenbürger* (Honorary Fellow) of the Vienna University of Technology, and in 2004 was awarded the *Austrian Cross of Honor for Arts and Science, First Class* from the President of the Republic of Austria. In 2005, he was inducted into the *Hunter College Alumni Hall of Fame*, and in 2006 he was awarded the prestigious *Townsend Harris Medal* by the City College Alumni

Association. He was inducted into the New York State *Mathematics Educator's Hall of Fame* in 2009, in 2010 he was awarded the coveted *Christian-Peter-Beuth Prize* in Berlin, and in 2017 he received the *Summa Cum Laude nemine discrepante* Award from Fundacion Sebastian, A.C. in Mexico City.

He has taken on numerous important leadership positions in mathematics education locally. He was a member of the New York State Education Commissioner's Blue-Ribbon Panel on the Math-A Regents Exams, and the Commissioner's Mathematics Standards Committee, which redefined the Standards for New York State, and he also served on the New York City schools' Chancellor's Math Advisory Panel. Dr. Posamentier is a leading commentator on educational issues and continues his long-time passion of seeking ways to make mathematics interesting to both teachers, students and the general public — as can be seen from some of his more recent books:

1. *Solving Problems in our Spatial World* (World Scientific, 2019).
2. *Tools to Help Your Children Learn Math*: *Strategies, Curiosities, and Stories to Make Math Fun for Parents and Children* (World Scientific, 2019).
3. *Math Makers*: *The Lives and Works of 50 Famous Mathematicians* (Prometheus, 2019).
4. *The Mathematics Coach Handbook* (World Scientific, 2019).
5. *The Mathematics of Everyday Life* (Prometheus, 2018).
6. *The Joy of Mathematics*: *Marvels, Novelties, And Neglected Gems That Are Rarely Taught in Math Class* (Prometheus, 2017).
7. *Strategy Games to Enhance Problem-Solving Ability in Mathematics* (World Scientific, 2017).
8. *The Circle*: *A Mathematical Exploration Beyond the Line* (Prometheus, 2016).
9. *Problem-Solving Strategies in Mathematics*: *From Common Approaches to Exemplary Strategies* (World Scientific, 2015).
10. *Effective Techniques to Motivate Mathematics Instruction* (Routledge, 2016).

11. *Numbers*: *Their Tales, Types and Treasures* (Prometheus, 2015).
12. *Teaching Secondary Mathematics*: *Techniques and Enrichment Units*, 9th Edition (Pearson, 2015).
13. *Mathematical Curiosities*: *A Treasure Trove of Unexpected Entertainments* (Prometheus, 2014).
14. *Geometry*: *Its Elements and Structure* (Dover, 2014).
15. *Magnificent Mistakes in Mathematics* (Prometheus Books, 2013).
16. *100 Commonly Asked Questions in Math Class*: *Answers that Promote Mathematical Understanding, Grades 6–12* (Corwin, 2013).
17. *What Successful Math Teachers Do*: *Grades 6–12* (Corwin, 2006, 2013).
18. *The Secrets of Triangles*: *A Mathematical Journey* (Prometheus Books, 2012).
19. *The Glorious Golden Ratio* (Prometheus Books, 2012).
20. *The Art of Motivating Students for Mathematics Instruction* (McGraw-Hill, 2011).
21. *The Pythagorean Theorem*: *Its Power and Glory* (Prometheus, 2010).
22. *Mathematical Amazements and Surprises*: *Fascinating Figures and Noteworthy Numbers* (Prometheus, 2009).
23. *Problem Solving in Mathematics*: *Grades 3–6*: *Powerful Strategies to Deepen Understanding* (Corwin, 2009).
24. *Problem-Solving Strategies for Efficient and Elegant Solutions*, *Grades 6–12* (Corwin, 2008).
25. *The Fabulous Fibonacci Numbers* (Prometheus Books, 2007).
26. *Progress in Mathematics K-9 Textbook Series* (Sadlier-Oxford, 2006–2009).
27. *What Successful Math Teacher Do*: *Grades K-5* (Corwin 2007).
28. *Exemplary Practices for Secondary Math Teachers* (ASCD, 2007).
29. *101+ Great Ideas to Introduce Key Concepts in Mathematics* (Corwin, 2006).
30. *π, A Biography of the World's Most Mysterious Number* (Prometheus Books, 2004).
31. *Math Wonders*: *To Inspire Teachers and Students* (ASCD, 2003).
32. *Math Charmers*: *Tantalizing Tidbits for the Mind* (Prometheus Books, 2003).

Gary Kose is Full Professor in the Psychology Department at Long Island University, Brooklyn Campus, and has been the Director of the M.A. Program in Psychology for the past 31 years.

Professor Kose received his B.A. degree from Temple University, in Philadelphia, PA, in psychology and philosophy in 1976. While there, he worked for two years in the Child Development and Learning Center, associated with the University of Pennsylvania Medical School. Upon graduation, he attended the Developmental Psychology Program at The Graduate Center, City University of New York. He completed his Ph.D. at The Graduate Center, as an NIMH fellow, in 1982. His graduate work concerned Piagetian theory and the development of representational capabilities in young children, with a special emphasis on the effects of media on the comprehension of pictorial representations. From 1981 to 1984, Professor Kose joined the faculty at the Institute for Cognitive Studies at Rutgers University. There his research interests broadened to encompass problems in general cognition, where he completed studies in memory, narrative comprehension, and problem solving looking at the effects of context and action. In 1984, he joined the faculty at Long Island University, Brooklyn Campus, teaching primarily in the doctoral program in Clinical Psychology. From 1988 to 1997, he served as Chair of the Department and Director of the M.A. Program in Psychology, a role which he maintains to the present. From 2002–2005, Professor Kose was Director of the Career Opportunities in Research Program, an NIMH-funded program designed to help undergraduate students pursue careers in research.

Throughout, Professor Kose has been an active member of a number of committees within the university and in professional organizations. Within the university, he has worked on the Graduate Goals Committee, the Middle States Accreditation Committee, and the Internal Review Board for the Brooklyn Campus. Professor Kose has been a long-time member of the Jean Piaget Society and the

International Society for Theoretical Psychology. In 2002, Professor Kose completed a summer fellowship program entitled Connecting Mind, Brain, and the Educational Institute at the Harvard Graduate School of Education. He has also been a member of the American Psychological Association, The Society for Research in Child Development, and the Psychonomic Society.

For the past 33 years, he has taught courses in the History of Psychology, Developmental Psychology, Cognition, and Research Design & Statistics. His research interests include cognitive development, Piagetian theory, theory of mind, cognition, problem solving, semiotics, and the psychology of the arts. Professor Kose has published over 40 articles in professional journals, book reviews, and book chapters. The following are a representative sample:

Corris, D. and Kose, G. (1997). The effects of action on imagery and memory. *Perceptual and Motor Skills*, **87**, 979–983.

Fireman, G. and Kose, G. (1990). Piaget, Vygotsky, and the development of consciousness. In Wm.J. Baker, M.E. Hyland, R. van Herewijk and S. Terwee (eds.), *Recent Trends in Theoretical Psychology*, New York: Springer-Verlag.

Fireman, G. and Kose, G. (2002). The development of self-regulation and awareness in children's problem solving performance. *Journal of Genetic Psychology*, **163**(4), 410–423.

Fireman, G., Kose, G. and Soloman, M. (2003). Self-observation and learning: The effects of watching oneself on performance. *Cognitive Development*, **18**, 339–354.

Fireman, G. and Kose, G. (2003). Psychotherapy: Science, myth, or both. In N. Stephenson, H.L. Radtke, R.J. Jorna and H.J. Stam (eds.), *Theoretical Psychology: Critical Contributions*. Concord, ON: Captus Press.

Fireman, G. and Kose, G. (2012). The development of perspective taking in young children. In E.H. Sandberg and B.L. Spritz (eds.), *The Clinician's Guide to Normal Cognitive Development in Childhood*. New York: Oxford University Press.

Kentgen, L., Allen, R., Kose, G. and Fong, R. (1998). The effects of representation of future performance. *British Journal of Developmental Psychology*, **16**, 505–517.

Kose, G., Beilin, H. and O'Connor, J. (1983). Children's comprehension of actions depicted in photographs. *Developmental Psychology*, **19**(4), 636–643.

Kose, G. (1984). The psychological investigation of art: Theoretical and methodological implications. In W.R. Crozier and A.J. Chapman (eds.), *Cognitive Processes in the Perception of Art*. Amsterdam: North-Holland.

Kose, G. (1985). Children's thinking about photography: A study of the developing awareness of a representational medium. *British Journal of Developmental Psychology*, **3**, 373–384.

Kose, G. (1987). A philosopher's conception of Piaget: Piagetian theory reconsidered [Review of the book *Beyond Piaget: A Philosophical Psychology*, J.P. Brief]. *Theoretical & Philosophical Psychology*, **7**(1), 52–57.

Kose, G. and Heindel, P. (1990). The effects of action and organization on children's memory. *Journal of Experimental Child Psychology*, **50**, 416–428.

Kose, G. (1992). Existential themes in Piaget's genetic epistemology. *Theory and Psychology*, **4**(2), 19–30.

Kose, G. and Corriss, D. (1994). Imaging: A theoretical alternative. In H.J. Stam, L.P. Mos, W. Thorngate and B. Kaplan (eds.), *Recent Trends in Theoretical Psychology*. New York: Springer-Verlag.

Kose, G. (1996). Piaget, born again! *Theory and Psychology*, **2**(3), 201–204.

Kose, G. (2002). The quest for self identity: Time, narrative, and the late prose of Samuel Beckett. *Journal of Constructivist Psychology*, **15**, 171–183.

Kosegarten, J. and Kose, G. (2010). Aspects of Wittgenstein's psychological concepts. In *Recent Trends in Theoretical Psychology*. Concord, ON: Captus Press.

Kosegarten, J., Kose, G. and Creedon, T. (2017). Metarepresentations reconsidered. In J. Cresswell, A. Haye, A. Larrain, M. Morgan and G. Sullivan (eds.), *Dialogue and Debate in the Making of Theoretical Psychology*. Concord, ON: Captus Press.

Kosegarten, J. and Kose, G. (2019). Logic, fast and slow: The persistent difficulty of the Monty Hall problem. *Journal of Evolutionary Psychology*, **2**, 50–72.

O'Connor, J., Beilin, H. and Kose, G. (1981). Children's belief in photographic fidelity. *Developmental Psychology*, **17**(6), 859–865.

Silvestri, H. and Kose, G. (2003). Rationality and the practices of science: A rereading of Kant. In N. Stephenson, H.L. Radtke, R.J. Jorna and H.J. Stam (eds.), *Theoretical Psychology: Critical Contributions*. Concord, ON: Captus Press.

Danielle Sauro Virgadamo is a clinical child psychologist whose clinical interests focus on parent training and behavioral problems in children. She received her B.A. in Mathematics and Psychology from The College of New Jersey in 2010, her M.S. in Applied Psychology from Long Island University — Post Campus in 2014, and her Psy.D. in Clinical Psychology from Long Island University — Post Campus in 2016.

Dr. Virgadamo has worked in various therapeutic settings, including private practice, outpatient clinics, inpatient units, and day treatment settings throughout New York and New Jersey. She completed psychology externships at Children's Specialized Hospital in Mountainside, NJ and Nassau University Medical Center in East Meadow, NY, where she specialized in work with children and their families. She completed her internship at Astor Services for Children and Families in the Bronx, where she worked with children with disruptive behavior disorders, anxiety disorders, and mood disorders. Dr. Virgadamo completed a two-year postdoctoral position at Cognitive Behavioral Associates, a private practice in Great Neck, NY, where she was foundationally trained in Dialectical Behavioral Therapy (DBT) and worked with adolescents with mood disorders and self-injurious behaviors. She also continued her work with young children and their parents, specifically using Parent–Child Interaction Therapy (PCIT) and other behavioral parent management strategies.

During her training at Long Island University, Dr. Virgadamo co-founded Family Check-In, a three-session assessment and referral program for underserved families with children aged 2–7. This program consists of a comprehensive evaluation of parent and child symptoms, a parent–child interaction followed by a reflection of the interaction, goal setting, and a feedback session that includes recommendations and referrals. She served as the Administrative Coordinator for Family Check-In and was responsible for creating clinician treatment manuals, phone screen form, chart checklists, scripts for each session, and feedback forms. Additionally, she was responsible for training first- and second-year students to implement the program in Long Island University's Psychological Services Center (PSC), the primary training clinic for second-year doctoral students. The program has continued its work with underserved families since its implementation and maintains a significant presence in the PSC. Dr. Virgadamo has presented numerous posters at national conferences regarding the implementation of Family Check-In and is currently writing an article about the feasibility and acceptability of the program that is to be submitted for publication.

After graduating from Long Island University, Dr. Virgadamo was invited to serve as adjunct faculty for both the undergraduate psychology department and the clinical psychology doctoral program. At the undergraduate level, she taught Introduction to Psychology (PSY1) and Principles of Psychology (PSY3), and in the doctoral program, she taught Advanced Statistics (PSY802).

Dr. Virgadamo's research interests focus on school-based mental health interventions, scale development, and mental health in twins. In 2012, she traveled to Kitengesa, Uganda, to assess children, interview caregivers, and collect data to determine the effects of a weekly library group on children's literacy, theory of mind, symbolic play, and school readiness. In 2013, she coauthored a chapter on school-based mental health programs, and in 2016 she coauthored a second chapter that compared school-based mental health interventions in the United States to those in Australia. In 2016, she developed and validated a scale that measures interpersonal interactions between twins. The measure, called the Sauro Twin Interpersonal Interactions (TwInI) Scale, determines the levels of dominance/submissiveness, competition/cooperation, and independence/dependence in adult twins.

In clinical practice, Dr. Virgadamo works primarily with children, adolescents, and their families. She specializes in parent training for children with disruptive and oppositional behaviors and additionally works with children and adolescents with anxiety and mood disorders. She has experience with psychological assessment, crisis management, family therapy, and group therapy. Dr. Virgadamo currently lives in Baltimore, MD, where she works as a clinical child psychologist at Kennedy Krieger Institute specializing in psychological assessment and child therapy.

Kathleen Keefe-Cooperman is a licensed clinical psychologist as well as an Associate Professor at Long Island University in the Department of Counseling and Development. Dr. Keefe-Cooperman has a B.A. in Psychology from Rhode Island College. She then went on to obtain an M.S. degree in Counseling from Pace University, and then another Master's degree in the area of Clinical Practices in Psychology from the University of Hartford, where she also received a Psy.D. in Clinical Psychology. She has been active in the fields of counseling and psychology in her work with students and professional peers and has been pursuing her research in the related fields as well.

Dr. Keefe-Cooperman has taught both civilians and military, as well as in the Army's West Point Tactical Officer's program. Watching graduate students pursue newly learned information encourages her to continue to pursue innovative areas of scholarly growth and share her findings with her classes as well as within the larger professional field. Dr. Keefe-Cooperman has been involved in research throughout her career. She first started while still in her doctoral program by developing annotated bibliographies in the area of improving patient–physician communication in the oncology field for a large nonprofit healthcare communication company. Dr. Keefe-Cooperman has since conducted research and published in the areas of perinatal loss, breaking bad news, and preschool adaptive functioning. Her research on the history of a state mental health hospital during the Great Depression resulted in her receiving an award. Her research has been frequently cited by others. A common theme throughout her research is the exploration of "when bad things happen to good people," and identifying barriers to success.

As one example of her work, Dr. Keefe-Cooperman developed the set communication protocol model "PEWTER" for counselors communicating bad news. There are a lack of structured communication protocols in the field of counseling which necessitated the need for this model. The mnemonic represents skills needed to facilitate difficult discussions. The "PEWTER" model has been well received. It was adopted by professionals involved in death notification for families of homicide victims, as well as those who must inform critical care patients of a loved one's death.

Dr. Keefe-Cooperman has provided psychological services on an ongoing basis to developmental treatment organizations. She is an expert in the area of preschooler development and has conducted countless psychological evaluations as part of a larger team. Dr. Keefe-Cooperman works closely with this age group. She previously conducted research looking at sleep patterns and adaptive functioning, finding that the effect of sleep is related to a variety of cognitive and behavioral patterns. Dr. Keefe-Cooperman also examined sleep patterns over time within the preschool population. Her expertise with this population caused her to note a difference in visual–spatial functioning following a cultural explosion in the area of digital device usage. Her research found a key relationship between a child's ability to engage in visual spatial functioning, such as completing puzzles, and smart device usage. Children who had more smart device usage also often had greater trouble with many fine motor tasks. Her published work in this area was well-received and led to her being awarded the prestigious Fulbright Specialist position. This involved sharing her work in Vienna as well as at top psychological conferences, both nationally and internationally.

Her presentation proposals are selected on a continuous basis for the highly respected American Psychological Association Convention. Dr. Keefe-Cooperman has also been involved in the area of teaching of psychology and was the past Chairperson for the American Psychological Association's Society for the Teaching of Psychology: Diversity Committee. She has worked to provide improved teaching strategies in the area of diversity.

Dr. Keefe-Cooperman is also an active volunteer and uses her expertise to provide a positive impact within communities. She served as a library trustee, as well as being involved with her local schools. Dr. Keefe-Cooperman uses her skills and knowledge to educate parents and children in the community about bullying and has also been interviewed on the topic for a documentary. Improving self-esteem in youth is the key to lessening the feelings of hopelessness often felt by children who are bullied. She is also involved with the Heroic Imagination Project, which teaches individuals about situational awareness, self-efficacy, values-driven decision-making, and social resilience.

Her ongoing work has set the stage for this book. Understanding the challenges people face that negatively impact their success is key to helping overcome those barriers. However, identifying the barriers is not enough. Dr. Keefe-Cooperman works with her fellow authors to provide the skills needed to learn successful ways to take on new problems with confidence. Individuals are not cemented into one mindset and can grow in order to attain success. The following are a representative sample:

Baker, L., Meiner, K. and Keefe-Cooperman, K. (2000). *Annotated bibliography for: Treating patients with C.A.R.E.* Institute for Healthcare Communication, Ontario, Canada (electronic version).

Brady-Amoon, M. and Keefe-Cooperman, K. (2017). Psychology, counseling psychology, and professional counseling: Shared roots; different professions? *The European Journal of Counselling Psychology*, **6**(1), 41–62.

Colangelo, J. and Keefe-Cooperman, K. (2012). Understanding the impact of childhood sexual abuse on women's sexuality. *Journal of Mental Health Counseling*, **34**, 14–37.

Keefe-Cooperman, K. (2005). A comparison of grief as related to miscarriage and termination for fetal abnormality. *Omega*, **50**(4), 281–300.

Keefe-Cooperman, K. (2016). Digital media and preschoolers: Implications for visual spatial development. *NHSA Dialog: The Research-to-Practice Journal for the Early Childhood Field*, **18**(4), 24–42.

Keefe-Cooperman, K. (2016). Preschooler digital usage and visual spatial performance: Implications for the classroom. *NHSA Dialog: The Research-to-Practice Journal for the Early Childhood Field*, **18**(4), 111–116.

Keefe-Cooperman, K. (2018). Training teachers in digital literacy. *Scientific Journal of PH Lower Austria*, **12**, 1–5.

Keefe-Cooperman, K. and Brady-Amoon, P. (2013). Breaking bad news in counseling: Applying the PEWTER model in the school setting. *Journal of Creativity in Mental Health*, **8**, 265–277.

Keefe-Cooperman, K. and Brady-Amoon, P. (2013). Preschoolers' sleep: Current U.S. community data within an historical and sociocultural context. *Journal of Early Childhood and Infant Psychology*, **8**, 35–55.

Keefe-Cooperman, K. and Brady-Amoon, M. (2014). Preschooler sleep patterns related to cognitive and adaptive functioning. *Early Education and Development*, **26**(6), 859–874.

Keefe-Cooperman, K., Savitsky, D., Koshel, W., Bhat, V. and Cooperman, J. (2017). The PEWTER study: Breaking bad news communication skills training for counseling programs. *International Journal for the Advancement of Counselling*, https://doi.org/10.1007/s10447-017-9313-z.

Keefe-Cooperman, K. L. (2000). *Annotated Bibliography for Improving Patient-Physician Communication in Oncology*. Bayer Institute for Health Care Communication, West Haven, Connecticut (electronic version).

Nardi, T. J. and Keefe-Cooperman, K. (2006). Communicating bad news: A model for emergency mental health helpers. *International Journal of Emergency Mental Health*, **8**(3), 203–207.

Contents

Introduction

Problem solving is a part of everyday life. We solve innumerable problems throughout our day, such as splitting the bill at a restaurant, finding the best route to avoid traffic when driving, ordering enough food for everyone at a party, arranging furniture, and decorating the house. For many people, these problems are manageable and solved without significant mental effort or anxiety. However, more complex situations involving multi-step problems inevitably exist, including planning and sticking to a schedule, meeting deadlines at work or school, and managing finances. Although some people excel at conquering such intricate problems, these situations are usually perceived as more challenging.

It is important to be skilled at problem solving in order to more easily tackle important problems that arise daily. Strong problem-solving skills are also regarded highly by others, from the time we are in school to the workplace and beyond. We are taught from the time we are young that mathematics and problem-solving skills are crucial. Many of us regard the two areas as important, but do not realize the innate similarities that make the skills more alike than different. Mathematics is a key component in problem solving. Young children who are labeled "good at math" are often assigned to a separate class and admired for their skill set. This separation rarely occurs for other subjects that young children study — sometimes reading, but hardly ever for social studies or science. In the workplace and throughout life, that message endures. People who are able to solve important and sometimes life-threatening problems are esteemed as "problem solvers," and "quick on their feet."

The strong emphasis placed on problem solving as a path to success in life might lead many to ask, "How can I become a better problem-solver?" Some people are naturally skilled at mathematics and problem solving and are confident in their abilities. Others have struggled with mathematics for their entire lives. Most people are somewhere in the middle. So where do these differences come from? Are we genetically predisposed to being good or bad at mathematics, and therefore, problem solving as well? Do genetics tell us if we will be confident or insecure in these abilities? Or is being good at problem solving a learned behavior? The truth is, we don't always have a clear picture as to why some people excel at mathematics and problem solving, and others struggle with it. Although one's intellectual ability is a contributing factor in problem solving, numerous psychological factors affect one's ability to solve problems, as well. Several factors that are crucial in problem solving are called executive functions, which are a set of complex processes that help people control behaviors and effectively solve problems. Several important executive functions are: attentional control, working memory, flexibility in thinking, cognitive inhibition (a form of self-control), and planning. Because strong executive functioning skills lead to better problem solving, which is such an essential part of life, it follows that if we want to become better problem-solvers, we should focus on strengthening these key executive functions.

You may have guessed that this book will target the executive functions listed above. In addition to executive functions, other psychological factors also affect our ability to solve problems. This book will first provide a brief history of problem solving and then target several of the most prevalent psychological factors that impede problem solving: misinterpreting or over complicating the problem, inflexibility, mathematics anxiety, inattention, forgetfulness, and impulsivity. We will review these factors as well as their counterparts: problem clarity and organization, cognitive flexibility, confidence, focused attention, working memory, and careful planning. This book will discuss how these cognitive strengths and weaknesses interact and affect our decision-making and problem solving. It will provide examples of common mathematical and everyday problems and the psychological factors that are interwoven within each problem to help the reader tap into the necessary principles.

This book is geared toward both readers who have the fixed belief that they are "bad at math" and those who are interested in refining their problem-solving skills by using psychological principles to train their brains. By explaining how key psychological factors can drastically affect our abilities to solve problems, this book will challenge the idea that people are born with a natural problem-solving ability. We hope to give confidence to those who struggle with mathematics and logic problems and offer a fresh perspective for mathematics and logic enthusiasts in approaching future problems they might encounter.

Chapter 1

A Brief History of Problem Solving

Humans are problem-solving animals. Aristotle defined humans as *Zoon Logikon*, which is loosely translated as the *rational animal*, although rationality was hardly defined at that time. And, what better evidence is there that we are rational animals than our long history of interest in games, mysteries, and puzzles throughout the evolution of human civilization. Beyond the sheer utility of being a good problem solver, humans have been fascinated and entertained with problem solving throughout recorded history and probably even before that point. Such activities generate suspense that calls for relief: a *carthesis* — the emotional relief that comes from watching a tragic drama, unraveling a great mystery, and solving challenging problems. One example of the history of humans' fascination with problem solving is the riddle of the Sphinx from the ancient legend of Oedipus. The Sphinx, a monster with the head and breast of a woman, the body of a lion, and wings of a bird, accosts all who dare enter Thebes by posing a riddle: "What is it that has four feet in morning, two feet at noon, and three at twilight?" Failure to solve the riddle results in death; the correct solution destroys the Sphinx. Oedipus answers, "Man, who crawls on four limbs as a baby (in the morning of life), walks upright on two as an adult (at the noon hour of life), and gets around with the aid of a stick in old age (at the twilight of life); thus, he unknowingly unlocks the riddle of his tragic past and future."

Or consider *King Minos*, who had a labyrinth built to hold the Minotaur — a monster half human and half bull — to avenge the death of his son by the Athenians. Fourteen Athenian youths were to enter the Labyrinth each year. None escaped until Thesus, with the help of Ariadne, King Minos's daughter, slew the Minotaur and found his way out of the twisting passages by simply following the thread he unwound himself on the way into the labyrinth. Such fantastic, mythic adventures capture not reality but rather accentuate the very real way emotion and reason come together in human experience.

Among the oldest books in recorded history are collections of games puzzles, and problems. One of the earliest surviving manuscripts, is a collection of mathematical puzzles known as the *Ahmes Papyrus*, named after the Egyptian scribe who copied it (also known as the *Rind Papyrus*, so named after the Scottish antiquarian who purchased it in 1858). The Papyrus is an anonymous work copied in 33 BCE, found in the ruins in a site near ancient Thebes. It contains 84 challenging arithmetic, geometric, and algebraic problems, with tables for calculating areas, the conversion of fractions, and linear equations as well as other information about measurement. The Papyrus begins with the epigram: "*Accurate reckoning, the entrance into the knowledge of all existing things and obscure secrets.*" It is believed that the Papyrus was intended as a practical manual to illustrate that accurate reckoning was the key to successful problem solving. Some of the problems have surfaced in other parts of the world and in languages unrelated to Ahems's. One problem (Number 79) presents an inventory without a suitable question:

Houses	7
Cats	49
Mice	343
Sheaves of wheat	2401
Hekats of grain	16,807
Estates	19,607

The inventory appears as a problem in the book *Liber Abaci* by the Italian mathematician Leonardo Pisano, today known as Fibonacci (*c.* 1170–1250), who adds a question and an additional power of seven:

> Seven old women are on the road to Rome. Each woman has seven mules, each mule carries seven sacks, each sack contains seven loaves, to slice each loaf there are seven knifes, and for each knife there are seven sheaths to hold it. How many are there altogether: women, mules, sacks, loaves, knife, and sheaths?

In this problem, there is a specific numerical concept in mind, the successive powers of seven: $7^1, 7^2, 7^3, 7^4, 7^5, 7^6$; while the last number is the sum of the previous numbers. Another version of the problem shows up in the eighteenth century English nursery rhyme, *Going to St. Ives*, but with a twist:

> As I was going to St. Ives,
> I met a man with seven wives,
> Each wife had seven sacks,
> Each sack had seven cats,
> Each cat had seven kits,
> Cats, kits, sacks, and wives,
> How many were going to St. Ives?

Here the trap is that the rhyme asks, "How many are going to St. Ives?" and not "How many are coming from St. Ives?" Only the narrator is going to St. Ives, all the others were coming from the city.

Many of the early collections were intended to educate. *Arithmetica* by the Alexandrian mathematician Diophantus (200–284 C.E.) was made up of puzzles to illustrate algebraic equations. The *Greek Anthology* is a compilation of riddles, epigrams, and mathematical puzzles similar to those in the Ahems Papyrus. Alcuin's (735–804 C.E.) *Problems for Sharpening the Young* (commissioned by Charlemagne) and Muḥammad ibn Mūsā al-Khwārizmī's (1048–1123 C.E.) *Calculation by Restoration and Reduction* are popular examples. By the thirteenth century such anthologies were common, among which the *Liber Abaci* (1202) by Fibonacci is the most famous. Fibonacci designed the book to be a pleasurable introduction to the Hindu–Arabic number system that about

50 years later became dominant throughout Europe. In the third section of the book (Chapter 12) is perhaps his most famous problem, the "rabbit" puzzle. Let's take a look at this trendsetting book.

The Book, *Liber Abaci*

Although Fibonacci wrote several books, the one for which he is best known is *Liber Abaci*. This extensive volume is full of very interesting problems. Based on the arithmetic and algebra, which Fibonacci had accumulated during his travels, it was widely copied and imitated, and, as noted earlier, the book introduced into Europe the Hindu–Arabic place-valued decimal system along with the use of Arabic numerals. The book was widely used for the better part of the next two centuries — a best seller!

Fibonacci begins *Liber Abaci* with the following paragraph, which is the first appearance of these numerals on the European continent.

"The nine Indian[1] figures are:

9 8 7 6 5 4 3 2 1.

With these nine figures, and with the sign 0, which the Arabs call zephyr, any number whatsoever is written, as demonstrated below. A number is a sum of units, and through the addition of them the number increases by steps without end. First one composes those numbers, which are from one to ten. Second, from the tens are made those numbers, which are from ten up to one hundred. Third, from the hundreds are made those numbers, which are from one hundred up to one thousand. ... and thus, by an unending sequence of steps, any number whatsoever is constructed by joining the preceding numbers. The first place in the writing of the numbers is at the right. The second follows the first to the left."

Despite their relative facility, these numerals were not widely accepted by merchants who were suspicious of those who knew how to use them. They were simply afraid of being cheated.

Interestingly, *Liber Abaci* also contains simultaneous linear equations. Many of the problems that Fibonacci considers, however, were similar to those appearing in Arab sources. This does not detract from the value of the book, since it is the collection of solutions to these problems that makes the major contribution to our development of mathematics. As a matter of fact, a number of mathematical terms — common in today's usage — were first introduced in *Liber Abaci*. Fibonacci referred to "factus ex multiplicatione"[a] and from this first sighting of the word, we speak of the "factors of a number" or the "factors of a multiplication." Another example of words whose introduction into the current mathematics vocabulary seems to stem from this famous book are the words "numerator" and "denominator."

The second section of *Liber Abaci* includes a large collection of problems aimed at merchants. They relate to the price of goods, how to convert between the various currencies in use in Mediterranean countries, calculate profit on transactions, and problems that had probably originated in China.

Fibonacci was aware of a merchant's desire to circumvent the church's ban on charging interest on loans. So, he devised a way to hide the interest in a higher initial sum than the actual loan, and based his calculations on compound interest.

The third section of the book contains a wide variety of problems:

A hound whose speed increases arithmetically chases a hare whose speed also increases arithmetically. How far do they travel before the hound catches the hare?

A spider climbs so many feet up a wall each day and slips back a fixed number each night. How many days does it take him to climb the wall?

Calculate the amount of money two people have after a certain amount changes hands and the proportional increase and decrease are given.

Some of the classical problems, which are considered recreational mathematics today, first appeared in the Western world in *Liber Abaci*. Yet the

---

[a] David Eugene Smith, *History of Mathematics*, Vol. 2, New York: Dover, 1958, p. 105.

technique to obtain a solution was always the chief concern for introducing the problem. This book is of interest to us, not only because it was the first publication in Western culture to use the Hindu numerals to replace the clumsy Roman numerals, or because Fibonacci was the first to use a horizontal fraction bar, but because it casually includes a recreational mathematics problem in Chapter 12 that has made Fibonacci famous for posterity. This is the problem on the regeneration of rabbits.

The Rabbit Problem

Figure 1.1 shows how the problem was stated (with the marginal note included).

A copy of the original page is shown in Figure 1.2.

To see how this problem would look on a monthly basis, consider the chart in Figure 1.3. If we assume that a pair of baby rabbits (B)

Beginning	1	"A certain man had one pair of rabbits together in a certain enclosed place, and one wishes to know how many are created from the pair in one year when it is the nature of them in a single month to bear another pair, and in the second month those born to bear also. Because the above written pair in the first month bore, you will double it; there will be two pairs in one month. One of these, namely the first, bears in the second month, and thus there are in the second month 3 pairs; of these in one month two are pregnant and in the third month 2 pairs of rabbits are born and thus there are 5 pairs in the month; in this month 3 pairs are pregnant and in the fourth month there are 8 pairs, of which 5 pairs bear another 5 pairs; these are added to the 8 pairs making 13 pairs in the fifth month; these 5 pairs that are born in this month do not mate in this month, but another 8 pairs are pregnant, and thus there are in the sixth month 21 pairs; to these are added the 13 pairs that are born in the seventh month; there will be 34 pairs in this month; to this are added the 21 pairs that are born in the eighth month; there will be 55 pairs in this month; to these are added the 34 pairs that are born in the ninth month; there will be 89 pairs in this month; to these are added again the 55 pairs that are both in the tenth month; there will be 144 pairs in this month; to these are added again the 89 pairs that are born in the eleventh month; there will be 233 pairs in this month. To these are still added the 144 pairs that are born in the last month; there will be 377 pairs and this many pairs are produced from the above-written pair in the mentioned place at the end of one year.
First	2	
Second	3	
Third	5	
Fourth	8	
Fifth	13	
Sixth	21	
Seventh	34	
Eighth	55	
Ninth	89	
Tenth	144	You can indeed see in the margin how we operated, namely that we added the first number to the second, namely the 1 to the 2, and the second to the third and the third to the fourth and the fourth to the fifth, and thus one after another until we added the tenth to the eleventh, namely the 144 to the 233, and we had the above-written sum of rabbits, namely 377 and thus you can in order find it for an unending number of months."
Eleventh	233	
Twelfth	377	

Figure 1.1

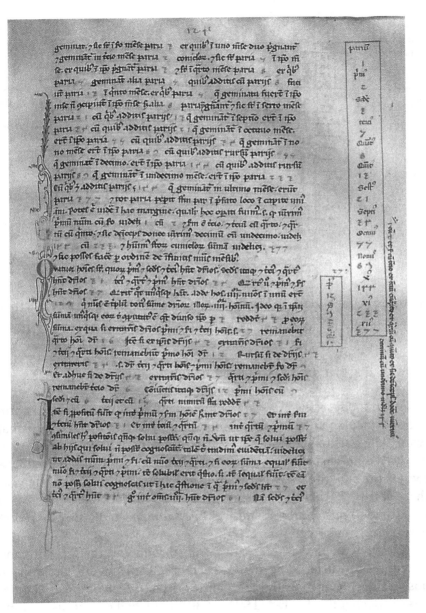

Figure 1.2

Month	Pairs	No. of Pairs of Adults (A)	No. of Pairs of Babies (B)	Total Pairs
Jan. 1	A	1	0	1
Feb. 1	A B	1	1	2
Mar. 1	A B A	2	1	3
Apr. 1	A B A A B	3	2	5
May 1	A B A A B A B A	5	3	8
June 1	A B A A B A B A	8	5	13
July 1	A B A A B A B A A B A A B	13	8	21
Aug. 1		21	13	34
Sept. 1		34	21	55
Oct. 1		55	34	89
Nov. 1		89	55	144
Dec. 1		144	89	233
Jan. 1		233	144	377

Figure 1.3

matures in one month to become offspring-producing adults (*A*), then we can set up the chart as in Figure 1.3.

This problem, which generated the sequence of numbers

1, 1, 2, 3, 5, 8, 13, 21, 34, 55, 89, 144, 233, 377, ... ,

is known today as the *Fibonacci numbers*. At first glance there is nothing spectacular about these numbers beyond the relationship that would allow us to generate additional numbers of the sequence quite easily. We notice that every number in the sequence (after the first two) is the sum of the two preceding numbers. At the end of one year there were 233 rabbits. Now known as the *Fibonacci sequence*, it can be extended *ad infinitum*, and has been unexpectedly discovered in nature and in many human activities.[2]

Puzzle making was a lucrative craft throughout the fifteenth and sixteenth centuries. For many, certain puzzles were thought to possess occult and aesthetic qualities. In 1612, in France, Claude-Gaspar Bachet de Mezirac (1581–1638) published one of the all-time bestselling collections, *Amusing and Delightful Number Problems*.

In 1779, the famous Swiss mathematician Leonhard Euler (1707–1783) introduced mathematical combinatoric in his *Thirty-Six Officers Puzzle*. The nineteenth century began with the British mathematician August De Morgan's (1806–1871) *A Budget of Paradoxes* (1826) and ended with Lewis Carroll's *Pillow Problems* (1880) and *Tangled Tales* (1885). In 1883, the French mathematician Francois Edouard Anatole Lucas (1842–1891), known for his writings in mathematics, and largely responsible for popularizing the Fibonacci numbers, invented the Towers of Hanoi Problem:

> A monastery in Hanoi has three pegs. One holds 64 gold discs in descending order or size — the largest at the bottom, the smallest at the top. The monks have orders from God to move all the discs to the third peg while keeping them in descending order. The three pegs can be used. When the monks move the last disk, the world will end. Why?

The world ends because it will take the monks $2^{64}-1$ moves to accomplish the task. At one move per second, with no mistakes, this task will require 582,000,000,000 years!

The twentieth century saw a proliferation of interest in puzzles of all kinds, including the invention of crossword puzzles in 1913 and the Rubik's Cube in 1974. Along with numerous societies and associations, a number of individuals, such as Martin Gardner (1914–2011), David Singmaster (1939–) and Raymond Smullyan (1919–2018), have helped establish recreational problem solving as a common pastime. This brief survey supports the idea that humans are, in fact, *rational animals*, and as Marcel Danesi (2002) contends, humans may possess a *"puzzle instinct."*[3] Yet, despite this impressive lineage of puzzle and problem collections, few if any of these texts discuss the psychological processes that underlie how we come to solve puzzles and problems. The Ahmes Papyrus mentions "accurate reckoning" as a process involved in problem solving. Beyond that, these collections serve to illustrate how problems and puzzles often conceal the answers and cry out for a solution.

The Psychological Aspects of Problem Solving

In the late nineteenth century, the discipline of psychology was established on scientific grounds within academies. Many of the psychological topics were derived from philosophical discussions about sensory perception. Wilhelm Wundt (1832–1920), at the University of Leipzig, set out to experimentally examine sensation and perception among other things. Wundt believed that conscious thought was based on sensory images; abstract and conceptual thinking was believed to be beyond the scope of experimental study. Oswald Kulpe (1862–1915), a former student of Wundt's, disagreed. Instead, he maintained that higher mental processes are imageless, yet could be demonstrated in a systematically, scientific manner through observations of problem solving. Some of Kulpe's most influential work came from his laboratory, when he was at the University of Wurzburg, on the *Einstellung* (attitude) or mental set effect. A very simple example of the *Einstellung* effect, used by Kulpe, was to present participants with cards with nonsense syllables written in different colors and arrangements. Some participants were instructed to focus on the colors, others to focus on the syllables. When asked to report what they saw, those instructed to focus on the colors recalled the colors quite well, but not so the syllables; conversely, those told to focus on the syllables did relatively well with the syllables, but not so with the colors. It appears that the instructions created a mental set directing attention to certain stimulus features and away from others, thus demonstrating that environmental sensory stimuli do not automatically create sensations that become conscious images or thoughts.

The most impressive study of the *Einstellung* effect, however, is the Water Jar problem, replicated and extended in a book *Rigidity of Behavior: A Variational Approach to Einstellung*[4] by Luchins and Luchins. In this study, participants were presented with a hypothetical situation in which they were given 3 jars of varying sizes and an unlimited supply of water, and were then asked to figure out how to obtain a specific required amount of water. The protocol and results from an experiment are presented below. The first item is a practice example (fill the 29 unit container and subtract 3 units from it).

The experimental group was given problems 2–11, in that order, to be solved by each participant; the control group was given the same introduction and practice problem but were instructed to begin working on problems 7–11. Problems 2–6 were called Einstellung problems, because they evoke the same problem-solving set $b - a - 2c$. Problems 7, 8, 10, and 11 are critical problems, because they could be solved by either a shorter more direct method ($a - c$, or $a + c$) or by the longer Einstellung method. Problem 9 is a recovery problem, to help the participants to recover from the Einstellung effect and have a fair chance of seeing the shorter solution for problems 10 and 11.

Problem	Given jars of the following sizes			Obtain the amount
	a	b	c	
1. Practice	29	3	—	26
2. Einstellung 1	21	127	3	100
3. Einstellung 2	14	163	25	99
4. Einstellung 3	18	43	10	5
5. Einstellung 4	9	42	6	21
6. Einstellung 5	20	59	4	31
7. Critical 1	23	49	3	20
8. Critical 2	15	39	3	18
9. Recovery	28	76	3	25
10. Critical 3	18	48	4	22
11. Critical 4	14	36	8	6

Below are the two types of possible answers for critical problems 7, 8, 10, 11

Problem	Einstellung solution	Direct solution
7	$49 - 23 - 3 - 3 = 20$	$23 - 3 = 20$
8	$39 - 15 - 3 - 3 = 18$	$15 + 3 = 18$
10	$48 - 18 - 4 - 4 = 22$	$18 + 4 = 22$
11	$36 - 14 - 8 - 8 = 6$	$14 - 8 = 6$

The following are the results showing the typical performance on critical problems:

Group	Einstellung solution	Direct solution
Control children	1%	89%
Experimental children	72%	24%
Control adults	0%	100%
Experimental adults	74%	26%

The results show that in the control group 89% of the children and 100% of the adults discovered the short, direct solution; whereas, in the experimental group only 24% of the children and 26% of the adults used the direct solution. In the experimental condition, 72% of the children and 74% of the adults used the longer Einstellung solution. It was concluded that Einstellung creates a mechanized state of mind or mental set that results in a blind attitude towards simpler, more obvious solutions. We do not look at a problem on its own merits, but are led by the mechanical application of a mental set that has been useful in other circumstances. This is often typical in the school curriculum when there are many textbook exercises that present problems that can create a mental set for a certain solution. This can prevent students from seeing new problems on their own merits and miss possible solution strategies.

Others have referred to this effect as *negative transfer*. Fredrick Bartlett[5] discovered this effect when observing his students solving an "alphametics" problem. The task in such a problem is to substitute numbers for the letters in the problem so that every number from 0 to 9 has its corresponding letter, and that each letter must be given a number different from any other letter, and so that the arithmetic (in this example, addition) is correct. Follow along as we solve one such alphametics problem:

The following letters represent the digits of a simple addition:

$$\begin{array}{r} \text{S E N D} \\ + \text{M O R E} \\ \hline \text{M O N E Y} \end{array}$$

We need to find the digits that represent the letters to make this addition correct. Most important in this activity is the analysis, and particular attention should be given to the reasoning used.

The sum of two four-digit numbers cannot yield a number greater than 19,999. Therefore **M = 1**.

We then have **MORE** < 2000 and **SEND** < 10,000. It follows that **MONEY** < 12,000. Thus, **O** can be either 0 or 1. But the 1 is already used; therefore, **O = 0**. We now have

$$
\begin{array}{r}
\text{S E N D} \\
\text{1 0 R E} \\
\hline
\text{1 0 N E Y}
\end{array}
$$

Now **MORE** < 1100. If **SEND** were less than 9000, then **MONEY** < 10,100, which would imply that **N = 0**. But this cannot be since 0 was already used; therefore **SEND** > 9000, so that **S = 9**. We now have

$$
\begin{array}{r}
\text{9 E N D} \\
\text{1 0 R E} \\
\hline
\text{1 0 N E Y}
\end{array}
$$

The remaining digits from which we may complete the problem are {2, 3, 4, 5, 6, 7, 8}. Let us examine the units digits. The greatest sum is $7 + 8 = 15$ and the least sum is $2 + 3 = 5$. If $D + E < 10$, then $D + E = Y$ with no carry over into the tens column. Otherwise $D + E = Y + 10$, with a 1 carried over to the tens column. Taking this argument one step further to the tens column, we get $N + R = E$, with no carryover, or $N + R = E + 10$, with a carryover of 1 to the hundreds column. However, if there is no carryover to the hundreds column, then $E + 0 = N$, which implies that $E = N$. This is not permissible. Therefore, there must be a carryover to the hundreds column. So, $N + R = E + 10$, and $E + 0 + 1 = N$, or $E + 1 = N$.

Substituting this value for **N** into the previous equation we get: $(E + 1) + R = E + 10$, which implies that $R = 9$. But this has already been used for the value of **S**. We must try a different approach. We shall assume, therefore, that $D + E = Y + 10$, since we apparently need a carryover into the tens column, where we just reached a dead end.

Now the sum in the tens column is $1 + 2 + 3 < 1 + N + R < 1 + 7 + 8$. If, however, $1 + N + R < 10$, there will be no carryover to the hundreds column, leaving the previous dilemma of $E = N$, which is not allowed. We then have $1 + N + R = E + 10$, which insures the needed carryover to the hundreds column. Therefore, $1 + E + 0 = N$, or $E + 1 = N$. Substituting this in the above equation $(1 + N + R = E + 10)$ gives us $1 + (E + 1) + R = E + 10$, or $R = 8$. We now have

$$\begin{array}{r} 9\ E\ N\ D \\ .\ \ \underline{1\ 0\ 8\ E} \\ 1\ 0\ N\ E\ Y \end{array}$$

From the remaining list of available digits, we find that $D + E < 14$. So, from the equation $D + E = Y + 10$, Y is either 2 or 3. If $Y = 3$, then $D + E = 13$, implying that the digits D and E can take on only 6 or 7. If $D = 6$ and $E = 7$, then from the previous equation $E + 1 = N$, we would have $N = 8$, which is unacceptable, since we already have $R = 8$. If $D = 7$ and $E = 8$, then from the previous equation $E + 1 = N$, we would have $N = 9$, which is unacceptable, since $S = 9$. Therefore, $Y = 2$. We now have

$$\begin{array}{r} 9\ E\ N\ D \\ .\ \ \underline{1\ 0\ 8\ E} \\ 1\ 0\ N\ E\ 2 \end{array}$$

Thus $D + E = 12$. The only way to get this sum is with 5 and 7.

If $E = 7$, we once again get from $E + 1 = N$, the contradictory $N = 8$, which is not acceptable. Therefore, $D = 7$ and $E = 5$. We can now again use the equation $E + 1 = N$ to get $N = 6$. Finally, we get the solution:

$$\begin{array}{r} 9\ 5\ 6\ 7 \\ .\ \ \underline{1\ 0\ 8\ 5} \\ 1\ 0\ 6\ 5\ 2 \end{array}$$

Bartlett found that students had the most difficulty due to their habit of solving addition and subtraction problems working from right to left. It seemed to Bartlett that the habit of solving to make an addition sum from the right-hand column and continuing to the left with succeeding columns was so deeply ingrained that they could not conceive of jumping to another pair of numbers out of sequence.

In a similar way, Karl Duncker[6] (1903–1945) investigated how past experience could limit problem-solving ability through his notion of *functional fixedness*, which was defined as a mental block against using an object in a *new way* that is required to solve a problem. Initially, individuals were presented with the tumor problem: "Given a human being with an inoperable stomach tumor, and rays which destroy organic tissue at sufficient intensity, by what procedure can one free him of the tumor with these rays and at the same time avoid destroying healthy tissue that surrounds it?" By examining the protocols of individuals attempting to solve this problem, it was found that problem-solving procedures move from general solutions to functional solutions to specific solutions. For example, a general solution might be "avoid contact between the rays and the healthy tissue;" once the individual thought of this, they hit upon several functional solutions, such as "use free path to the stomach," or "remove healthy tissue from the path," and then reach specific solutions, such as "use esophagus" or "insert a cannula." If these solutions failed, new general and functional solutions were proposed, such as "lower intensity of the rays on their way through the healthy tissue"; more specific ideas followed, "turn down the intensity when the rays get near the healthy tissue and turn it up when it gets to the tumor" and then finally to a correct solution, "use focused lenses" to converge on the tumor (see Figure 1.4). According to Duncker, the problem is a result of fixation on the idea of a single ray; the solution came as a sudden flash of insight, an "Aha" experience, and the recognition that the ray could be divided into weaker rays that can converge on the tumor.

To investigate this phenomenon more carefully, Duncker devised a series of problems that he thought involved overcoming functional fixedness. For example, he presented participants with three cardboard boxes, matches, thumbtacks, and candles. The goal was to mount the candle vertically on a nearby wall to serve as a lamp. Some of the subjects were given a box containing matches, a second box holding candles, and a third box containing tacks (i.e., using the boxes

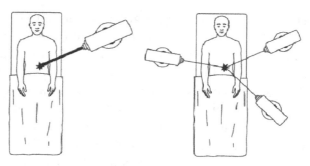

Figure 1.4

(a) Preutilization condition   (b) No preutilization condition   (c) Solution

Figure 1.5

as containers); while other subjects were given the same supplies with the matches, candles, and tacks outside the boxes (see Figure 1.5).

The solution — to mount a candle on top of a box by melting wax onto the box and sticking the candle to it and then tacking the box to the wall — was much harder to discover when the boxes were given filled rather than empty. Duncker's explanation was that the placement of the objects inside the box helped to fix the function as a container, thus making it more difficult to reformulate the function of the box as a support. Duncker examined variations of this problem using paper clips and string; all confirmed that functional fixedness is an obstacle to effective problem solving. Together, the work on the Einstellung effect, negative transfer, and functional fixedness provided evidence that the reapplication of very rigid, past habits may hinder productive problem solving.

The Gestalt Approach to Problem Solving

While these early studies demonstrate important obstacles to problem solving, they do not provide a clear idea of the positive aspects of *efficient* problem solving. How do we solve problems? According to Gestalt (form) psychologists, problem solving is a process of searching to relate each aspect of a problem situation to each other in order to achieve a structural understanding; that is, to comprehend how all the parts of the problem fit together to satisfy the requirements of the goal, which involves reorganizing the elements of the problem in a new way.

An important aspect of this process is overcoming a pre-established problem-solving set that limits how the problem is perceived. Since we cannot look at the problem situation in a new way, we cannot see new ways of re-structuring or fitting the elements together. For example, when trying to solve the nine-dot problem (Figure 1.6), which requires us to connect all the dots with four straight lines, and without lifting one's pencil from the paper, finding a solution is often very

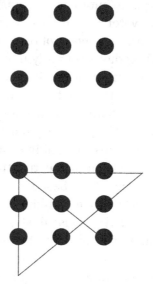

Figure 1.6

difficult. The nine dots appear to form a square that imposes a problem-solving set (that the solution must be contained within the square). Solutions can be found when you consider that the lines can extend beyond the perimeter of the apparent square.

This approach was first proposed by the German psychologist Wolfgang Köhler[7] (1887–1967), a founder of Gestalt psychology. Köhler spent four years on the island of Tenerife in the Atlantic Ocean, where he made extensive observations of primates' problem-solving behavior. He published his findings in a monograph, *The Mentality of Apes*. In a typical situation, he observed an ape named, Sultan, in a large penned area that contained crates and sticks, and bananas hanging from the ceiling out of reach. Kohler watched as Sultan unsuccessfully attempted to reach the bananas. After a long pause, described as intense thinking (incubation), there followed what appeared to be a flash of *insight* — Sultan placed the crates one on top of another and climbed the crates to reach the bananas. The solution, according to Kohler, was not the result of continuous trial and error, but rather the result of a reorganization of the elements in the problem situation after a period of incubation and insight.

It is hard not to imagine Sultan as a primitive Archimedes, when in his tub watching the water rise, he discovered how to determine the volume of the king's crown; it was not reported if Sultan attempted to grunt "Eureka!" While today, we hardly use such vague, imprecise terms as insight, and reorganization, it is clear that this Gestalt approach is trying to get at the creative mental process needed for problem solving with novel solutions. In his book *Productive Thinking,* Max Wertheimer[8] (1880–1943), the original founder of the Gestalt School, distinguished two kinds of thinking. One involved creating a new solution to a problem called productive thinking because a new organization is produced; the other, based on applying past solutions to a problem, called reproductive thinking because old habits are simply applied. This distinction was illustrated by suggesting two methods of teaching students how to find the area of a parallelogram. One method emphasized the geometrical or structural properties of the parallelogram — for example, showing how the triangle on one

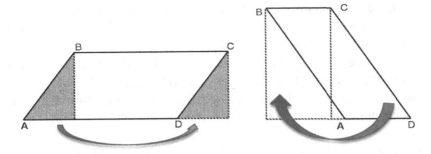

Figure 1.7

end of the figure could be placed on the other end, thus forming a rectangle (see left side of Figure 1.7). The other method emphasized a sort of recipe of steps to calculate the area by dropping the perpendicular and multiplying height times the length of the base.

Although students taught by both methods performed equally well on the criterion tasks that involved finding the area of parallelograms like those they had learned about, they differed in their ability to transfer what they had learned to new tasks. The students who learned by "understanding the structural properties" were able to find the areas of unusual parallelograms and shapes, and to recognize incalculable situations; while the students who learned the mechanical receipt had difficulty extending what they had learned, often responding with, "We haven't learned that yet."

This Gestalt approach to solving problems has led to attempts to break the thinking process down into several distinct stages. As early as 1926, the English social psychologist Graham Wallis (1858–1932) argued that creative problem solving proceeds through four stages: In the first stage, *preparation*, the problem solver gathers information about the problem. This stage is often characterized by hard and frustrating work on the problem, generally with little progress. In the second stage, *incubation*, the problem solver sets the problem aside and seems not to be working on it; although he or she may be working on a solution unconsciously ("like a baby bird develops unseen, inside the egg"). The period of incubation leads to the third stage, *illumination*, in which some key insight or new idea emerges. This stage paves the way for the fourth stage, *verification*, in which the problem solver confirms the possible solution.

More recently, the well-known Hungarian mathematician George Polya (1887–1985) offered a series of steps in problem solving in his book *How to Solve It.*[9] These steps are based on observations he made of his students while he was a mathematics teacher.

(1) *Understanding the problem* — the solver gathers information about the problem and asks, "What do you want (or what is unknown)? What have you (or what are the data and conditions)?"

(2) *Devising a plan* — the solver gathers information about the problem and asks, "Do I know a related problem? Can I restate the goal in a new way based on my past experience (working backwards) or Can I restate the givens in a new way that relates to my past experience (working forward)?"

(3) *Carrying out the plan* — the solver tries out the plan of the solution, checking each step.

(4) *Looking back* — the solver tries to check the result by using another method, or by seeing how it all fits together, and asks, "Can I use this result or method for other problems?"

Polya illustrated the steps with the following geometry problem:

Find the volume of the frustum of a right pyramid with a square base given the altitude *h* of the frustum, the length of a side of *area* 1 of its lower base, and the length of a side of its upper base *area* 2 (Figure 1.8).

The Solution Process

(1) *Understanding the problem.* Solver asks: What do I want? Answer: The volume of the frustum. Solver asks: What do you have? Answer: Length of a side of area 1, length of a side of area 2, and the height.

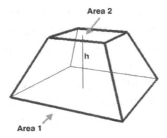

Area 2

h

Area 1

Figure 1.8

(2) *Devising a plan.* If you cannot solve the proposed problem, look around for an appropriate-related problem. Solver asks: What is a related problem that I can solve? Answer: The volume of a full pyramid. Solver asks, Can I restate the goal of the given problem differently? Answer: Restate the goal as the volume of the full pyramid minus the volume of the smaller pyramid in the upper portion. Restate the givens to yield the height of the full pyramid and the height of the smaller pyramid in the upper portion.

(3) *Carry out the plan.* Use known formulas to find the height of the full pyramid, the height of the small pyramid, and the volume of each. Solution: Subtract the volume of the smaller pyramid from the volume of the larger pyramid. $(V = \frac{1}{3}hB)$.

(4) *Look back.* Solver asks: Can I use this method for other problems? Do I see the overall logic of this method?

While Wallis and Polya's stages may seem no more than a description of what occurs during problem solving, they do help us see more clearly what is involved in the Gestalt idea of insight and restructuring. Establishing and restating a goal, finding a plan by working forward or backward from the goal, and carrying out and checking the plan is a clearer account of what is involved in restructuring and solving a problem. So, this might be what Sultan was doing while staring and "intensely thinking" about those bananas!

The Gestalt psychologists' attempt to understand the complex mental processes behind problem solving has introduced several provocative ideas. They underscore the rigidity in thinking and the effects of a mental set as a possible obstacle to problem solving. The distinction between productive and reproductive thinking helps clarify the complexity of thinking that is involved in solving problems. Further, thinking about finding a solution may be best thought of in stages: understanding the problem, restructuring or reorganizing the elements of a problem, and testing the solution. In this process, there is reason to consider periods of incubation and insight.

Further Thoughts on Creative Problem Solving

If restructuring is crucial, how do problem solvers go about seeking some new perspective, some new mental set, when working on a difficult problem? An alternate way to approach this issue is by examining cases in which someone finds an entirely new and wonderfully productive way to solve a problem. We can do this by examining cases of individuals who have been undeniably creative — artists like Pablo Picasso, Ludwig van Beethoven, Richard Wagner, or innovative scientists like Charles Darwin and Marie Curie. Studies have indicated that these creative individuals share things in common. These individuals all have an enormous storehouse of knowledge and skill in their domain of achievement. They are often intelligent, not afraid to take risks, are willing to ignore criticism, and have the ability to tolerate ambiguous findings or situations. Also, highly creative people are motivated by the pleasure of their work rather than the promise of external reward. Finally, it has been noted that such creative people have been "in the right place at the right time" — that is, in environments that provide freedom, support, and problems that are ripe for solutions with the available resources. But these various factors merely set the stage of creativity; we still need to ask what goes on inside the mind when a creative person is seeking out a novel solution. What, if anything is different between people who are good problem solvers and others who are not? What is it that makes the difference?

There are many different perspectives on this point. Some stress innate factors, others stress social or cultural factors. But there is one factor that is crucial, and everyone seems to agree on: experience working in a particular domain. Thus, the quality of a physician's problem solving (making accurate diagnoses) is improved after years of medical practice; the quality of an electrician's problem solving (trouble shooting problems and laying complex wiring patterns) is enhanced by on the job experience. The same is true of painters, professors, or police officers — they all become better problem solvers with experience. But why exactly does experience improve problem solving?

Having experience is no guarantee of improving problem-solving skills; but if you are a good problem solver or consider yourself an expert, you will have considerable experience in your field of achievement. One of the reasons for this is that over the years of experience, good problem solvers gather a wealth of information about their domain of expertise. Indeed, this is why it is often suggested that someone usually needs a full decade to acquire expert status in a field, whether in mathematics, music, chess, or software design. It is estimated that 10 years of time is needed to have access to the necessary facts and have them cross-referenced.

It is important to be aware, though, good problem solvers don't just have more facts or information than novices; they have a different type of knowledge that is focused on higher-order patterns. This kind of knowledge enables experts to think in larger units, tackling problems in bigger steps rather than smaller ones. This ability is evident, for example, in studies of chess players. Novice chess players think about a game in terms of the position of individual pieces; experts, by contrast, think about the board in terms of pieces organized into strategic groups. This sort of thinking is made possible because the better chess players have a "chess vocabulary" in which complex concepts are stored as single units with an associated set of subroutines for how one should respond to an emerging pattern. There are investigators who estimate that some chess masters may have as many as 50,000 high-order chunks in memory, each representing a strategic pattern. These patterns can be detected in how players recall a game and in their sensitivity to problem organization. Also, knowledge of higher-order patterns helps good problem solvers and experts in terms of using analogies in their thinking. Experts routinely use analogies, giving them a great advantage in their work. Using analogies can be promoted by attention to the underlying structure of a problem rather than to surface characteristics of a problem.

Conclusion

Humans may be *rational animals*, yet we are far from flawless in our thinking and often have difficulty solving problems. We are prey to many limitations, such as incorrect conclusions, unsolved problems, and foolish decisions. The psychology of problem solving has shown that we are routinely victims of our own assumptions — although without those assumptions, many problems would be ill defined and more difficult. And, this raises the question as to whether it is possible to lead people towards better thinking that involves more accurate conclusions, more compelling deductions, better decisions, and more successful problem solving. Strategies can be helpful, and methods of education can be useful. Indeed, one often hears the call for education programs that promote critical thinking. Some of the most recent studies in thinking and problem solving have contributed to understanding strategies in problem-solving work, and how thinking can be improved.

Chapter 2

Exploring the Problem Space: Problem-Solving Strategies

In life, we are surrounded by many types of problems. Some are big, like deciding on a career or what kind of car to buy; we clearly know these are problems and consciously struggle with solutions. Other problems are practical: what to wear; how to cross the street, or where to eat tonight? Such problems we hardly recognized as problematic until our typical solution doesn't work (Americans often recall that after a late-night flight to London, walking out of a hotel and crossing a large street, looking first to the right, then to the left, but quickly realizing the need for a new solution to the typical, everyday problem of crossing the street in the United States or the European continent). Then there are specific problems that challenge us, like those we encounter in math classes and other educational situations. These problems often seem artificial or contrived to help us learn a routine or skill for finding solutions. And, then, sometimes we entertain ourselves with problems like trying to do a crossword puzzle or playing Sudoku. But before we talk about finding solutions, we need to define the different types of problems we encounter throughout our lives.

In psychology we make an important distinction between *ill-defined* and *well-defined* problems.[10] With ill-defined problems, you start the problem with only a vague sense of the goal. For example, imagine you are planning a "summer vacation." While you know you want something fun to do, there are many things you don't know: Stay close to home or travel? How much can I spend? Seaside, the mountains or travel to a city? Ill-defined problems provide only loose guidelines for problem solving. Thus, it is common that when we try to solve ill-defined problems, we start by making them well-defined — by clarifying and specifying the goal state. This often involves adding extra constraints (spend around $1,000.00) and assumptions (near the ocean). Of course, by narrowing the options, we may end up neglecting better options. Nonetheless, defining the problem more clearly helps enormously in a search for a solution.

Well-defined problems, on the other hand, are problems in which you have a clear idea of the goal right at the start, and you know what options are available to you in reaching that goal. For example, in solving the anagram (words with letters scrambled): *subufoal* — you know immediately the goal involves eight letters forming an English word; you also know that you'll reach that goal by rearranging the sequence of letters and not, for example, by turning letters upside down. You won't waste time adding letters or changing letters to numbers; the font and color of the ink are irrelevant for the goal. Within this problem space, we can hit on a solution: *fabulous*! With well-defined problems, you have a clear idea of the *problem space* — that is, your thoughts in solving a well-defined problem are guided by where you are and where you want to be. Understanding this *problem space*, between the *initial state* and the *goal state*, has become an important approach to the psychology of problem solving. With well-defined problems, you have a clear idea of the goal right at the start, and you know what options are available to you in reaching that goal. While many of us, when faced with a challenging problem, immediately try to think of a solution; good problem solvers and experts spend a lot of their time trying to understand the problem in term of its problem space. Finding a solution depends on how we understand and search through the problem space.

Algorithms

There are many different strategies for searching through a problem space that can be generally divided into *algorithms* and *heuristics*.[11] Algorithms are step-by-step procedures that will always produce a solution to a problem. Algorithms are often used in computers because of the rigid step-by-step procedures that are involved and the certainty of reaching a solution. In our everyday lives, algorithms are often automatic and can be described as habits (like when crossing the street and automatically looking to the right and the left). Heuristics, on the other hand, are rules of thumb that involve conducting a selective search, examining a portion of the problem space that is most likely to produce a solution. We will examine the concept of heuristics in greater detail later in this chapter as well as in the next chapter.

It is in our nature, when faced with a novel situation or problem, to approach it in terms of our past experience. Prior experience is the basis for much of our behavior and actions. When faced with a new problem, if we can see similarity with prior experience, we will use that experience to approach a solution, selecting an algorithm or heuristic that matches the problem. If we are correct, and the match fits, then the problem is easily solved. If, however, there is a mismatch between our prior experience and the problem we are facing, then we experience difficulty and frustration. In fact, our prior experience can blind us to clues in the new problem and even to obvious solutions. Today, this effect of prior experience is known as the *mental set effect* (or *Einstellung Effect* — see Chapter 1). It was found that after successfully using a solution strategy on several occasions, the strategy will be automatically repeated on other problems, even for those problems that are not appropriate and where a more obvious solution is available.

If we see the opportunity to use an algorithm, we must recognize the series of steps needed to solve a particular problem. That is, we have to see that there are steps that must be carried out in a particular order and the steps must not be replaced. Like a recipe, the algorithm tells you exactly what to do to solve a problem. As a very simple example, consider the following: you are given a square piece of metal measuring 8 feet × 8 feet, and a smaller square measuring 3 feet × 3 feet is cut out and discarded. The problem is to find the area of the remaining piece of metal. An algorithm for finding a solution could be as follows:

Draw five empty calculation boxes:

1.	Label the first box "large"	LARGE	8
2.	Label the second box "small"	SMALL	3
3.	Label the third box "sum"	SUM	11
4.	Label the fourth box "difference"	DIFFERENCE	5
5.	Label the fifth box "result"	RESULT	55

Then execute the following:

1. Record the size of one side of the large sheet of metal in the LARGE box;

2. Record the size of one side of the small sheet of metal in the SMALL box;

3. Calculate the sum of the LARGE box and the SMALL box and record the result in the SUM box;

4. Calculate the difference of the LARGE box and the SMALL box and record in the DIFFERENCE box;

5. Calculate the product of the SUM and the DIFFERENCE boxes and record the product in the RESULT box.

Using such an algorithm to solve this problem is most efficient. It does not require that you know how to find the area of a square; and, if you follow each step exactly, you will always arrive at a solution to this particular kind of problem. Sometimes, we learn such recipes or strategies in educational settings, such as a math class, and the rigid precision makes such algorithms prefect for being programmed in to a computer. Using algorithms can provide a quick and easy, positive experience for someone who is just learning how to solve problems. There are, however, certain limitations that make the use of algorithms not the best or optimal problem-solving strategy.

Algorithms can be easy to learn and quick to use and you must know exactly what strategic procedure to follow given a particular problem. You have to know exactly "what to do" given a certain problem; you don't have to know "why" you are doing it. You do not have to understand why the procedure works. This limitation affects the ways in which an algorithm can be used. Algorithms are developed for particular problems, and they do not work when there are slight variations in the problems we may be facing (e.g., the algorithm above does not work if you remove a rectangle instead of a square from the larger sheet) or when facing a completely novel problem. That is, we have difficulty transferring the knowledge of a strategy to variations of the same type of problems. Educators often experience this difficulty when children and adults fail to transfer strategic knowledge. For example, high school students are taught an algorithm to solve physics problems:

> What is the acceleration (increase in speed each second) of a train, if its speed increases uniformly from 15 miles/second at the beginning of the first second, to 45 miles/second at the end of the twelfth second?

While the students were effective at solving physics problems, they had little success transferring the algorithm to solving algebra problems:

> Juanita went to work as a teller in a bank at a salary of $12,400 per year and received constant yearly increases, reaching a $16,000 salary during her 13th year of work. What was her yearly salary increase?

One may have failed to see that the algorithm used with physics problems was relevant to solving algebra problems, which had the same structure as the physics problems. Algorithms give a specific receipt or procedure for solving a particular problem.[12] But when the strategy or meaning behind them is not understood, algorithms do not transfer to other problem situations, and because of this limitation there has been greater interest in the more useful heuristic strategies.

Heuristics

A heuristic strategy (derived from the ancient Greek meaning to "find" or "discover") is any approach to problem solving that uses a practical method, not guaranteed to be optimal, perfect, logical, or rational, but instead emphasizes the discovery of a sufficient way to reach a problem goal. The concept was originally introduced in Psychology by the American Nobel Laureate Herbert A. Simon (1916–2001). Simon's primary research showed that we typically approach problem solving within what he calls "bounded rationality." Heuristics are rules of thumb that suggest a strategy for working through a problem space. The most fundamental heuristic is *trial and error*, which can be used for everything from matching nuts and bolts, completing jigsaw puzzles, and finding the values of variables in algebra problems. Also, *exhaustive search*, where you explore every possible move or step to a solution, is another very basic heuristic. However, neither approach is very efficient. One of the most effective heuristic strategies is the *means-end analysis*. This strategy involves asking, "What's the difference between my current state and my goal?" Then with the difference defined, you ask, "What means do I have for reducing the difference?" For example, "I want to get to the store. What's the distance between where I am now and the store? How can I reduce the distance? By car, but my car won't work. What's needed to get it working? A battery, etc." A means-end analysis can replace the initial problem (getting to the store) with a series of sub-problems (getting the car to work, getting a battery). If this process is repeated, in that subgoals are broken down into smaller subgoals, a path forms to the desired solution.

Understanding the problem space in this way has several advantages. First, subproblems are likely to be less complex than the original problem, so they are easier to solve. Also, subgoals are often more straightforward. For example, a driver wanting to get to the store can realize that the best route is to take the highway, so "getting to the store" gets replaced with "get to the highway," which may be composed of subroutines that are easier to carry out. Lastly, having smaller subgoals makes the problem seem more manageable and not as overwhelming.

A great deal of research has been conducted on the *means-end heuristic* by studying how we come to find a solution to the *Tower of Hanoi* problem (see Figure 2.1). With this problem you are presented with a board that has three pegs. On the first peg there are three disks of different sizes in ascending order, with the largest on the bottom and the smallest on the top (see step 0). The goal is to move the disks one at a time so that you end up with the three disks stacked in the same order on the far end of the board (see step 7). A disk can be moved onto another peg as long as it is placed on a disk that is larger than the moving disk. There are many solution paths, and the most efficient seven-step solution is shown in Figure 2.1.[13]

The Tower of Hanoi problem has been described as a near "perfect problem" to study, because it has the key features of a well-defined problem. It has a clearly defined *initial state*, a clearly defined

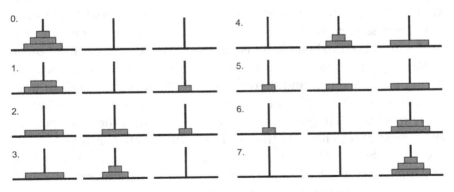

Figure 2.1

goal state, and a clear set of operators or moves for getting through the *problem space*. Also, it is "lean" in that achieving a solution does not require any specific knowledge or facts. Adults and most young children can solve the easy three disk problem with a little effort and thought. But the problem can be made more difficult by simply adding a disk; that is, with four disks the most efficient solution requires 15 steps. Most children and many adults have difficulty with a four-disk problem.

Most of us start solving the three-disk problem by *trial and error* or attempting to move directly to the goal by placing the smallest disk on the third peg (step 1). But quickly you realize you must consider the goal state more carefully. After the second move (step 2), moving the middle disk to the second peg, you encounter a new problem; you need to find a place for the largest disk. This is the first subgoal to solve. We could move the middle disk back to the first peg or move the smallest disk back to the first peg, but neither move resolves this subgoal — you still cannot move the largest disk. Both moves result in having to make many more moves to get to the final goal. The optimal move is to move the smallest disk back to the middle peg, placing it on top of the middle disk (step 3). Now the largest disk can be moved to the third peg (step 4). Usually, at this point the path to the final goal state is clearly seen.

At the heart of this means-end analysis is the process of establishing *subgoals*. When a move is made and an obstacle arises, the first subgoal is to remove the obstacle as the new desired goal. This three-step process is recursive in that it is repeated when a new obstacle arises until the final end state is reached. With the three-disk Tower of Hanoi the first obstacle occurs at step 2 where the disks are spread across the three pegs and there appears to be no further moves possible towards the final goal. The only possible moves are backward away from the final goal. Solving this subgoal frees the third peg for the largest disk to move to the third peg. Often, between step 3 and 4, there may be an *"Aha"* experience — children and adults sometimes exclaim, "Now I got it!" and the final three steps are accomplished very quickly with the goal clearly in sight.

Another popular problem for studying means-end analysis is the *Hobbits-and-Orcs* problem[14]:

> There are three hobbits and three orcs on one side of the river. Your goal is to ensure that the hobbits and orcs all arrive safely on the other side of the river by transporting them across the river in a boat. The boat can carry up to two creatures at a time. Orcs eat the hobbits, so the number of orcs cannot exceed the number of hobbits on either side of the river. What sequence of crossings will transport all six creatures across the river without causing harm to the hobbits?

This problem is more difficult.

The solution requires 12 steps, shown in Figure 2.2 (where the underline stands for the river, H = hobbits, O = orcs, b = boat). The first important subgoal is to realize that you must get all the hobbits on the other side of the river before you can safely move the orcs across. The difficulty in finding a solution is in step 6, in which you must send back the boat with both a hobbit and an orc. This step is counterintuitive, because it goes against our intuition to reduce the greatest difference between the current state and the goal state. In trying to solve this problem, most of us back up to step 5 (sending back two hobbits) rather than moving on to step 7 (sending back a hobbit and an orc). The Hobbit-and-Orc problem is difficult because we are reluctant to move backwards temporarily from the desired goal state.

Another helpful strategic heuristic, somewhat related to mean-ends analysis, is the *working-backward* heuristic. This strategy is commonly used when there are too many possible operators or moves that can be applied to the initial state of the problem. You start at the desired goal state and examine the possible conditions that would have to be true if the desired end was true. This process is continued for each step "prior-to-the-goal" state (you would have to determine again which states would be needed for the current state to be true). In the case of the three-disk Tower of Hanoi, this would require that you recognize that before the end state can be achieved, the largest disk has to be placed on the third peg; in order for this

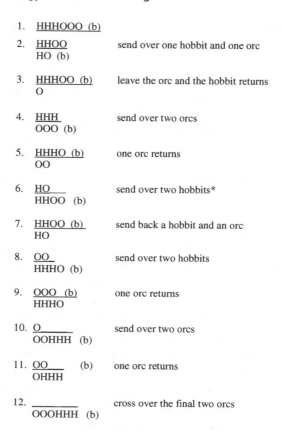

1. <u>HHHOOO (b)</u>

2. <u>HHOO</u>   send over one hobbit and one orc
   HO (b)

3. <u>HHHOO (b)</u>   leave the orc and the hobbit returns
   O

4. <u>HHH</u>   send over two orcs
   OOO (b)

5. <u>HHHO (b)</u>   one orc returns
   OO

6. <u>HO</u>   send over two hobbits*
   HHOO (b)

7. <u>HHOO (b)</u>   send back a hobbit and an orc
   HO

8. <u>OO</u>   send over two hobbits
   HHHO (b)

9. <u>OOO (b)</u>   one orc returns
   HHHO

10. <u>O_____</u>   send over two orcs
    OOHHH (b)

11. <u>OO___</u> (b)   one orc returns
    OHHH

12. <u>_____</u>   cross over the final two orcs
    OOOHHH (b)

Figure 2.2

to be achieved, the two smaller disks need to be placed on the second peg (see steps 3 and 4), and so on.

Consider as another example of working backwards the proof that the base angles of an isosceles triangle are equal (see Figure 2.3). Working backward would require asking how we can prove two angles equal. If these angles were corresponding parts of congruent triangles, then we would be able to conclude their equality. We then ask how we can create two congruent triangles from the given isosceles triangle. We could construct an angle bisector of the vertex angle *C*, which would give us two triangles, and if these two triangles are congruent, then we would be at our desired conclusion. We can create

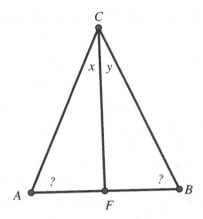

Figure 2.3

two possibly congruent triangles by constructing the angle bisector CF, such that $\angle x = \angle y$. Now $\triangle ACF \cong \triangle BCF$ (by the side-angle-side postulate). Therefore, since the triangles are congruent, and our desired angles are corresponding angles, we can conclude that $\angle A = \angle B$. The solution is an example of working backward, because to assume base angles are equal, we look for conditions that would have to be true if the desired end state was true.

Another more complex example that shows the power of using a working-backward technique is the following problem: The sum of two numbers is 2. The product of these same two numbers is 5. Find the sum of the reciprocals of these two numbers. A problem of this sort would not be uncommon for an algebra class. Most students would immediately begin by forming two equations in two variables:

$$x + y = 2,$$

$$xy = 5.$$

These two equations can be solved simultaneously by using the algorithm of the quadratic formula, which is $x = \frac{-b \pm \sqrt{b^2 - 4ac}}{2a}$, for $ax^2 + bx + c = 0$. However, the method yields complex values for both x and y, namely, $1 + 2i$, and $1 - 2i$. Following the requirements of the original problem, we now need to take the reciprocals of these two roots and find their sum:

$$\frac{1}{1+2i}+\frac{1}{1-2i}=\frac{(1-2i)+(1+2i)}{(1+2i)(1-2i)}=\frac{2}{5}.$$

We should emphasize here that there is nothing wrong with this method, it is just not the most elegant way to solve this problem, nor is it the easiest to comprehend.

In working backwards, we can step back from the problem and see what is being required. Curiously, this problem is not asking for the values of x and y, but rather the sum of the reciprocals of these two numbers. That is, we seek to find $\frac{1}{x}+\frac{1}{y}$. Using a strategy of working backwards, we could ask ourselves from where this might have come. Adding these two fractions could give us this answer. Therefore, $\frac{1}{x}+\frac{1}{y}=\frac{x+y}{xy}$. At this point the required answer is immediately available to us, since we know the sum of the numbers is 2, and the product of the numbers is 5. We merely substitute these values in the last fraction to $\frac{1}{x}+\frac{1}{y}=\frac{x+y}{xy}=\frac{2}{5}$, and our problem is solved.

Another problem that demonstrates the power of this working backwards procedure is the following: Lauren has an 11-liter can and a 5-liter can. How can she measure out exactly 7 liters of water? Most people will simply guess at the answer, and keep "pouring" back and forth in an attempt to arrive at the correct answer, a sort of "unintelligent" guessing and testing. However, the problem can be solved in a more organized manner by using the strategy of working backwards. We need to end up with 7 liters in the 11-liter can, leaving a total of 4 empty liters in the can. But, where do 4 empty liters come from (see Figure 2.4)?

To obtain 4 liters, we must leave 1 liter in the 5-liter can. Now, how can we obtain 1 liter in the 5-liter can? Fill the 11-liter can and pour from it twice into the 5-liter can, discarding the water. This leaves 1 liter in the 11-liter can. Pour the 1 liter into the 5-liter can (see Figure 2.5).

Now, fill the 11-liter can and pour off the 4 liters needed to fill the 5-liter can. This leaves the required 7 liters in the 11-liter can (see Figure 2.6).

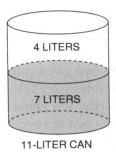

Figure 2.4

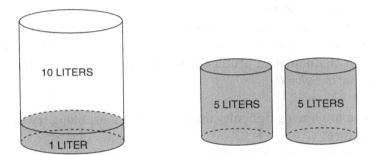

Figure 2.5

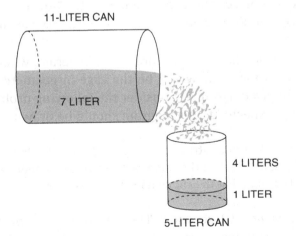

Figure 2.6

Note that problems of this sort do not always have a solution. That is, if you wish to construct additional problems of this sort, you must bear in mind that a solution only exists when the difference of multiples of the capacities of the two given cans can be made equal to the desired quantity. In this problem, $2(11) - 3(5) = 7$. This concept can lead to a discussion of parity. We know that the sum of two like parities will always be even (that is, even + even = even; odd + odd = even) whereas the sum of two unlike parities will always be odd (odd + even = odd). Thus, if two even quantities are given, they can never yield an odd quantity. Further discussion can be particularly fruitful as it gives the students a much-needed insight into some valuable number properties and concepts.

There are many different kinds of heuristics that can be used to solve a variety of problems. Here are a few other commonly used heuristics for mathematics problems, from George Polya's 1945 classic book, *How to Solve It*[15]:

- If you are having difficulty understanding a problem, try drawing a picture.
- If you can't find a solution, try assuming that you have a solution and seeing what you can derive from that ("working backward").
- If the problem is abstract, try examining a concrete example.
- Try solving a more general problem first (the "inventor's paradox": the more ambitious plan may have more chances of success).

As an example of what is meant drawing a diagram, where one isn't generally called for — as would be the case in geometry, which, of course, requires a diagram — consider the following problem, one in which it is *not* expected that a diagram should be drawn.

> At 5:00 o'clock, a clock strikes 5 chimes in 5 seconds. How long will it take the same clock at the same rate to strike 10 chimes at 10:00 o'clock? (Assume that the chime itself takes no time)

The answer is *not* 10 seconds! The nature of this problem does not lead us to anticipate that a drawing should be made. However,

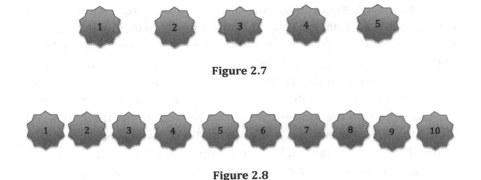

Figure 2.7

Figure 2.8

let us use a drawing of the situation to see exactly what is taking place. In the drawing, each dot represents a chime. Thus, in Figure 2.7, the total time is 5 seconds and there are 4 intervals between chimes.

Therefore, each interval must take $\frac{5}{4}$ seconds. Now let's examine the second case:

Here, we can see from the diagram in Figure 2.8 that the 10 chimes give us 9 intervals. Since each interval takes $\frac{5}{4}$ seconds, the entire clock striking at 10:00 o'clock will take $9 \cdot \frac{5}{4} = 11\frac{1}{4}$ seconds.

Other heuristics include adopting a different perspective on a problem (called *blockbusting*, which involves actively trying to look at a problem in a new way). Sometimes this can be accomplished by looking at problems in the extreme. Consider the following example of a problem whose solution can be best solved by considering an extreme case:

> We have two 1-liter bottles. One contains a half-liter of red wine and the other contains a half-liter of white wine. We take a spoonful of the red wine and pour it into the white-wine bottle, and thoroughly mix the two-colored wines. Then we take a spoon of this new mixture (red wine and white wine) and pour it into the red-wine bottle. Is there more red wine in the white-wine bottle, or more white wine in the red-wine bottle?

There are several common approaches, where the problem solver attempts to solve the problem using the given information — namely the spoon. With some luck, and cleverness, a correct solution may evolve, but it will not be easy and often not convincing. We can see that the size of the spoon does not really matter, since there are large and small spoons. Suppose we use a very large spoon, one that is enormously large and actually can hold half-liter of liquid — this would be an extreme consideration. When we pour the half-liter of the red wine into the white-wine bottle, the mixture is then 50% red wine and 50% white wine. After mixing these two together, we take our half-liter spoon take half the quantity of this red-wine — white-wine mixture and pour it back into the red-wine bottle. The mixture is now the same in both bottles; so, to answer our question we can conclude that there is as much red wine in the white-wine bottle as there is white wine in the red-wine bottle.

We can consider another form of using an extreme case, where the spoon doing the wine transporting has a zero quantity. In this case the conclusion follows immediately: There is as much red wine in the white-wine bottle as there is white wine in the red-wine bottle, that is, zero!

Yet another solution to this problem uses some logical reasoning. With the first "transport" of juice there is only red wine on the spoon. On the second "transport" of juice, there is as much white wine on the spoon as there is red wine in the "white-wine bottle." This may require a bit more thought, but most should "get it" soon.

But most importantly, blockbusting involves becoming aware of our fixation on how we initially see a problem and actively work to represent it in a new way. Many problems in geometry are solved by using logical (deductive) reasoning. Reorganizing the details or elements of a particular problem is also an effective heuristic. No one heuristic works for all problems; and, there is no guarantee that a heuristic will result in a solution. Heuristics can best be thought of as *tools* to help us achieve solutions. But the key to using any heuristic is discovery. You must discover the solution path for yourself. In this way, problems become ways of learning or teaching solution strategies. This becomes difficult in typical educational curricula. Discovering a heuristic requires time to explore the problem space. The emphasis is not so much on a correct solution, but on a correct path to a solution. Errors and false starts must be tolerated and used as teaching or learning moments. Very often a successful heuristic comes as a reaction to an error, or false start.

The Importance of Problem Representation

When faced with a problem, selecting the right strategy or tool for reaching a solution is a matter of how we represent or understand the problem space. Consider this insight problem:

> A man who lived in a small town married 20 women in the town. All the women are still living, and he has never divorced any of them. Yet the man has broken no laws. How is this situation possible?

The solution is that the man has broken no laws, because he is a priest. To arrive at this solution, you must represent the givens of the problem correctly. Most of us think of getting married in a passive sense, in that one "gets married"; the solution hinges on understanding the meaning of "married" and "divorce" as active, something that someone actively does "for others." You can see that in this case, using heuristics alone to attempt to solve the problem will not help you. You must reframe the problem space to create a new understanding of the problem.

A wonderful example of the importance of representation comes from a famous legend about the great German mathematician, Carl Friedrich Gauss (1777–1855).[16] Born in Brunswick, Germany, Gauss displayed immense mathematical talent from a very young age. Stories tell of him being able to maintain his father's business accounts at the age of three. In elementary school, he amazed his teacher by observing a pattern that allowed him to avoid a tedious calculation. Gauss's teacher asked the class to add together all the numbers from 1 to 100. Presumably the teacher's aim was to keep the students occupied for a time while he was engaged in something else. Unfortunately for the teacher, Gauss quickly spotted a shortcut to the solution. Gauss realized that rather than adding the numbers in the sequence provided (1+2+3+...+98+99+100), it would be more efficient if he added the first and the last numbers, $1 + 100 = 101$, the second number and the next-to-last number $2 + 99 = 101$, the third number and the third from last number, $3 + 98 = 101$, and continuing on in this fashion he would have 50 pairs of numbers totaling 101, therefore the sum required is simply $50 \times 101 = 5050$.

While Gauss's classmates plodded along, step-by-step — using what today would be called the *hill-climbing* strategy — proceeding towards the goal state, solving each new subgoal as it arose to reduce the difference between the initial state and the goal state:

$$10 + 5 = 15 \ldots$$
$$6 + 4 = 10$$
$$3 + 3 = 6$$
$$1 + 2 = 3$$

Gauss, instead, represented the problem in terms of 50 pairs — each summing to 101; giving him 5050.

Essentially, there is a regrouping in Gauss's solution — the reorganization of the series in light of the problem — where there is an attempt to grasp the inner relations between the sum of the series and its structure. As a result, various elements take on new meaning that are functionally determined by the goal. When Austro-Hungarian psychologist, Max Wertheimer (1880–1943) asked children of various ages to solve the same kind of problem (add all the numbers from 1 to 10) "without using the cumbersome additions," he found that few could match young Gauss. He did find, however, that eventually, a few children came to approximate Gauss's elegant solution. In some cases, children began to appreciate that the problem represents a series $(n+1)$, then that the series can be seen in terms of pairs $\left(\frac{n}{2}\right)$; some older children began to understand that the numbers can be seen in terms of ascending and descending series that are functionally related. Some children attempted to solve the problem by writing out the ascending series, then the descending series directly underneath, which allowed them to see that the sum of the pairs from the ascending and the descending series were all the same. While these solutions were not as elegant compared to Gauss, Wertheimer saw this as evidence that Gauss's solution was not merely a fortuitous stroke of genius, but a result of understanding the functional relationship within the structure of the problem; that is, seeing how the problem is represented.

A final, dramatic illustration of how important the representation of the problem is to finding a solution comes from a small study of anagram solving.[17] Undergraduate students were timed when solving anagram problems (words with letters scrambled). For example:

kmli   graus   teews   recma   foefce   ikrdn

Some students were shown lists of such word problems in an unorganized fashion and were timed in reaching a solution (the median time to solution was 12.2 seconds — but not the one shown above, whose solution is: *milk, sugar, sweet, cream, coffee, drink*). Others were shown such scrambled words that were organized around a certain theme — and the theme was represented early in the list — for example, *café* (with café in mind go back and try unscrambling the words). When the theme was recognized the students solved the same anagrams in about half the time (median time to completion 7.4 seconds). Representing or thinking about the anagrams in terms of a meaningful category quite literally affects how we see the elements and letters jump out at you forming words related to café (*milk, sugar, sweet, cream, coffee, drink*).

When facing a problem, we cannot help but use prior experience as a possible guide for finding a solution. But prior experience can either help or hinder us. Having a mental set can either prepare us to quickly use a prior solution or it can blind us to features of the novel problem and even obvious solutions. What makes the difference is how we represent the problem in light of past experience. As mentioned earlier, understanding the meaning of the word "marriage" in an active sense allows us to understand how a man can perform multiple marriages without breaking the law; or instead of seeing a problem as a series of addition problems, we can see it as the relationship between ascending and descending series; or instead of guessing at the word meaning of a jumble of letters, seeing the letters in terms of an organizing concept makes all the difference in finding solutions. How we represent the problem affects how we see the elements and details of the problem, and consequently how we search for a solution. How do we come up with different ways of representing various kinds of problems?

One way, which almost everyone agrees with, is to have experience with numerous types of problems and problem-solving strategies, and to practice using them. To be a good problem solver, you have to spend a good deal of time solving various types of problems. We would expect that having a repertoire of such problem-solving experiences would begin to show itself in the way we approach problems and ultimately solve them. Enough practice of this kind should, for the most part, make a longer-range goal attainable, namely, that we would naturally come to use these same problem-solving strategies in solving new problems that we encounter. This transfer of learning (back and forth) can be best realized by introducing problem-solving strategies in a variety of ways (e.g., mathematical, word problems, and real-life situations).

There are at least three ways to acquire problem strategies: we can be told about them, we can observe someone else use them, or we can discover them for ourselves.[18] Acquiring strategies by observing someone else using them or by being instructed how to use them involves aspects of social learning[19] and can be equally effective; however, an instruction along with observing an example is often the best method. It has been shown many times that providing a worked example is one of the most effective methods of instruction for developing problem-solving skills, and more formal explanations can be of great help when examples may obscure a strategy. For example, when children are shown a problem like, $3 \times 2 + 5 = 6 + 5 = 11$, and are then given $4 + 6 \times 2 = ?$; the children often incorrectly answer 20 (mistakenly adding 4 and 6, then multiplying that by 2). Instructions can alert them that they should always perform multiplication (and division) first in any such expression.

Discovering a problem-solving strategy for oneself is the preferred method of acquisition. Acquiring a strategy by discovery ensures that you will better understand the functioning of the strategy and, also, help with the transfer to variations of other problem situations. But discovering strategies for yourself can be complicated. Such discoveries are not always obvious; they can be time consuming, and can be frustrating. Educational settings are not always conducive to self-discovery. First, and perhaps foremost, we must have the right attitude towards problem solving.

Motivation

Differences in motivation are always evident.[20] In any given class, you might present the following simple cryptarithmatic problem and see two general reactions.

Try substituting numbers for the letters to make this work out arithmetically correct:

$$
\begin{array}{r}
AD \\
+\,D\,I \\
\hline
D\,I\,D
\end{array}
$$

Some students will take out paper and a pencil and pour over every letter. Others will quickly look at the problem, saying something like: "I'm not very good at things like this (or math)" or "I'm not interested in such things." The first group *engages* the problem — getting the solution:

$$
\begin{array}{r}
91 \\
+\,10 \\
\hline
101
\end{array}
$$

The latter group dismisses the problem. This distinction, while an obvious oversimplification, points to an important difference in motivational attitude. Those students who dismiss the problem are probably reflecting a history of failure or anxiety over being evaluated, and consequently dismiss the problem situation. They are not even trying. They have not failed to solve the problem; they have not even engaged the problem. Those students who engage the problem are

not threatened by failure or being evaluated, and have a greater tolerance of errors. Anxiety and motivation will be discussed in greater
detail in Chapter 4, particularly techniques for managing anxiety and
increasing motivation to become a better problem solver. Frustration
tolerance is also an important part of problem solving. For instance,
having a low frustration tolerance will impact the ability to persevere.
Thus, a prerequisite to discovering a problem-solving strategy is to
have a positive attitude towards problem solving that allows you to
intimately engage problems. Problem solving should not be seen as a
test to evaluate but should be seen as a way to teach or learn about
different ways to reach solutions while remaining calm.

Another prerequisite to discovering problem-solving strategies is
to change our attitude about the time it takes to reach a solution.
Often, we think that a good problem solver, or an intelligent person,
will arrive at a solution quickly. But speed can come when we have
already established a repertoire of strategies. Gauss probably came
so quickly to a solution because he already possessed a unique way
of thinking about a series of numbers. But when we first encounter
a particular type of problem, we need time to explore the problem
space; we need time to attempt certain strategies, make errors and
mistakes, and then try other approaches. In other words, we need to
approach the discovery of problem strategies in a noncompetitive
way and without becoming frustrated. This can be done at the school
level by having students work together on finding problem solutions,
rather than to compete with each other. Working together with others
has several other advantages, such as encouraging each other to ver-
balize thinking processes and to entertain different perspectives on a
problem. Students should also be praised on the problem-solving
process itself rather than just the solution.

Also, we have all had the experience of working on a problem,
feeling stuck and frustrated, and then leaving the problem for a while
to return with a new approach. Sometimes problem-solving strategies
emerge after a period of *incubation*. You're working on a problem, but
getting nowhere. After a while, you give up and turn your thoughts
to other matters. Then sometime later, when you're thinking about
something altogether different, a solution to the problem you were

working on pops into your thoughts. An example is someone waking up out of a sound sleep at 2:00 AM with the solution present in his or her thoughts. In the history of science there are many examples of discovering solution strategies in this way. Incubation probably works for a variety of reasons. Time away from a problem may allow our minds to access new clues and perspectives on a problem; or incubation may simply relieve fatigue; or incubation may help us break our initial fixation on how we first saw the problem, allowing for a more productive re-structuring of the problem. But at base, incubation removes the competitive edge and the idea that speed to a correct solution is important.

In sum, there is no easy answer when it comes to how we come to represent a problem space and acquire problem-solving strategies. The quickest and most efficient way to acquire strategies is through imitation and direct instruction. Although, acquiring strategies in this way is often limited in scope and application. Discovering strategies for oneself is the preferred method of acquisition — ensuring better understanding of the dynamics of the strategy and extending the range of application. Although, acquiring strategies by discovery can be time consuming, there is no guarantee that a strategy will be discovered, and guidance may be required.

Conclusion

In this chapter we have tried to approach problem solving through a variety of strategies as a topic in itself, and not focus on problems of any particular type. We hope you gained greater insight about your own problem-solving strategies and entertained new approaches to try in the future. Problem solving is no more or less than a search, regardless of the content or domain of a problem, albeit math, word problems, or real-life problems. Unfortunately, when faced with a problem, instead of searching for a solution, we often expect to see an answer pop out at us. Finding a solution requires a search through a problem space — comparing the distance between the initial state of the problem and the goal. In *How to Solve It*, George Polya summarized the general orientation to problem solving as follows: understand the problem, plan a solution, execute the plan, and check your results. If you are a poor problem solver you probably skip the first step and immediately look for a solution to the problem. Poor problem solvers often use trial and error in which the first solution that comes to mind is put into play. Better problem solvers, and experts, spend a good deal of time developing a full understanding of the problem space — comparing the initial state of the problem with the goal state.

Problem-solving strategies are the ways in which we traverse the problem space. There is no one strategy, or set of strategies, that is effective for all problems. Algorithms or heuristics are the two types of strategies we are likely to employ. The strategies we use are the result of how we represent or understand the problem space. There are, however, some strategies that are commonly used and have been thoroughly studied. The means-end heuristic is perhaps the most general strategy we use. At the core of this strategy is the act of establishing subgoals or subroutines. There is no easy way to acquire problem strategies — imitation, instruction, and discovery are all effective, but each has limitations. Although, there is one crucial factor for acquiring problem-solving strategies that everyone agrees upon: experience with a variety of problems is necessary. To be a good problem solver, solve as many different kinds of problem as you can. Such experience allows us to gather a large base of information about different problems and can help us see patterns across problems. But, of course, openness to problem-solving situations is contingent on overcoming the motivational and emotional obstacles to problem solving.

Chapter 3

Judgment, Reasoning, and Decisions

The activity of thinking takes many forms. To start, people often draw judgments based on their experienes — judgments about a person's personality, where to have dinner, or how the weather will be for an upcoming vacation? Can we count on these judgments? How exactly — and how well — do we learn from past experience? Then, once we form judgments, we take another step and think through the implications of our new beliefs. How do we go beyond the information gained from our experienes? Finally, we make countless decisions everyday based on our beliefs about our past experience. Many are trivial, such as having soup or salad? Others can change our lives in important ways — where to go to school, whether to marry, etc. In this chapter, we will consider the psychological aspects of forming *judgments*, and *reasoning* from our beliefs, and making *decisions*.

Heuristics

We first learned about heuristics in Chapter 2, and the concept is worth revisiting. Experience is a great teacher, and so we put considerable faith in the judgment that a physician makes on the basis of her years of experience, or in the advice from an auto mechanic based on the many cars he has worked on. But surely there are limits to what we can learn from experience. Sometimes, information from past experience is incomplete or ambiguous. Sometimes, our memories of our experiences are selective or even distorted. How do these considerations affect our ability to make judgments and draw conclusions about what we have seen, heard, or read?

For example, suppose you want to decide how difficult your school's biology course is. No doubt, you'll ask yourself, "How well have my friends done in this course? How many received good grades, and how many did poorly?" Here, we're thinking about frequencies. *Frequency estimation* is crucial for making judgments. However, it's likely that you have not kept an ongoing tally of your friend's grades, or if you have, you may be able to recall only a few who have taken the course. What most of us will do is to fall back on an *attribute substitution* strategy, using easily available information that is a substitute for the information we seek. In this case, you are likely to do a quick scan through memory looking for a relevant case. If you can think of two or three friends who got good grades, you'll conclude that this is a frequent occurrence and that the course is not challenging. If you have a hard time thinking of anyone with a good grade in the course, or if the only ones who come to mind had poor grades, you draw the opposite conclusion. With this strategy, you are basing your judgment on *availability*, that is, how easily or quickly you can come up with relevant examples. Treversky and Kahneman have called this the *availability heuristic*.[21] As mentioned earlier, heuristics are efficient strategies that usually lead to correct judgments. With the availability heuristic, the attribute being used is easily accessed — the attribute being relied on is correlated with our goal, so it can serve as a proxy for the target. Events and objects that are frequent are, in fact, likely to be easily available in memory and, so generally, you can rely on availability as an index for frequency. Nonetheless, this heuristic strategy can lead to errors. Let's take an example from the psychological literature: ask yourself, "Are there more words in the dictionary that begin with the letter 'R' (such as rose, rock, rabbit) or more words with 'R' in the third position (such as tarp, bare, throw)?" Most people answer that there are more words beginning with R but, in fact, the reverse is true. Why do so many of us get this wrong? The answer lies in availability.

If you search your memory for words beginning with R, many come to mind; not so for words with R in the third position. The difference favoring words beginning with R arises because memory is organized roughly like a dictionary, with words sharing the initial sound being grouped together. As a consequence, it is easier to search memory by using the initial sound of a word; a search for words with R in the third position is much more difficult. Thus, the organization of memory creates a bias in what is easily available, and this bias can lead to an error in frequency judgments.

While this experimental example is not very interesting on its own, it does illustrate very common mistakes we make in judgments. People regularly overestimate events that are quite rare. This probably plays a part in people's willingness to buy lottery tickets because they overestimate the probability of winning. Physicians often overestimate the likelihood of a rare disease and in the process fail to examine other more appropriate diagnoses. Why does this happen? There is little reason to think about familiar events ("Oh, look that plane is overhead in the sky."), but we are more likely to notice rare events, especially rare, emotional events ("Oh my gosh, that plane has crashed!"). As a result, rare events are likely to be well recorded in memory and, in turn, make these events more easily available to us. We then overestimate the frequency of these distinctive events and, correspondingly, overestimate the likelihood of similar events happening in the future.

A very nice study that illustrates the usefulness of the availability heuristic involved asking participants to think of episodes in their lives in which they acted in an assertive way.[22] Half of the participants were asked to recall six such episodes; the other half were asked to recall 12 episodes. Then all participants were asked some general questions, including to assess how assertive overall they thought they were. Participants had an easy time thinking of six episodes, using the availability heuristic; in contrast, participants who were asked for 12 episodes had difficulty generating the longer list. Consistent with the functioning of the availability heuristic, participants who easily recalled fewer episodes judged themselves to be more assertive. Ironically, participants who recalled more episodes, with difficulty, actually reported themselves as less assertive. This half had more evidence of assertiveness, but it was not the quantity of evidence that mattered; instead, what mattered was the ease of coming up with the episodes. The participants who were asked for a dozen episodes had a hard time with the task because they had been asked to do something difficult, namely, come up with a lot of cases. But the participants seemed not to realize this. They reacted only to the fact that the examples were difficult to generate and, using the availability heuristic, concluded that being assertive was relatively infrequent in their past.

As a final example, imagine that you've been asked to vote on how much money the government should spend on research projects, all aimed at saving lives. Obviously, we should choose to spend resources on projects that address the more frequent causes of death. Given this logic, should we spend more on preventing death from stomach cancer or motor vehicle accidents? Should we spend more on preventing homicides or diabetes? More people reliably assent that motor vehicle accidents and homicides are more frequent and deserve more money; although, the opposite is true. Death from stomach cancer and diabetes are more frequent. Here, we can appreciate the effect of the media. While stomach cancer and diabetes are more frequent causes of death, they get less coverage. While auto accident deaths and homicides are actually less frequent, they are likely to be covered in the media, and thus seem to be more frequent, because they are available to influence our judgments.

As you can see, at times, the availability heuristic can be beneficial, and at other times, it can trick our brains into believing false information. So, how do we make sure that we use the availability heuristic in the best possible way, particularly when solving problems? Unfortunately, it can be hard to tell when a belief is formed based on evidence or simply the information to which we are exposed. However, remaining practically skeptical and keeping yourself informed on topics of interest is a good start. In Chapter 5, we will discuss a strategy to help you remain nonjudgmental, which will help stop you from making snap judgments. When all else fails, do research yourself to determine the actual frequencies of an event instead of relying on word of mouth. Remind yourself that the media (including social media) tends to focus more on *uncommon* events; however, when we see many of these rare events, we believe them to be much more common than they actually are.

Another way to understand how we come to make judgments is our use of the *representative heuristic.*[23] Consider that you are applying for a job. You hope the employer will give full consideration to your past experience and qualifications. The employer, however, may rely on a more efficient strategy — one that involves another type of attribute substitution. The employer may barely glance at your resume and ask himself, instead, how much you resemble other people he previously hired who worked out well. Do you have the same looks and mannerisms as another employee that he was very happy with? If you do, you are likely to get the job. In this case, the employer is making a probability judgment based on resemblance. Just like the availability heuristic, this strategy is efficient and can lead to correct conclusions. But, just like the availability heuristic, this strategy can lead you astray. The representative heuristic is based on the assumption that all members of a category all pretty much look alike and are the same: all birds have wings, feathers, and can fly; all offices have desks, chairs, and computers. As a result, we use resemblance for judging who or what belongs to a category — "what *looks* alike *is* alike." While this type of reasoning often leads to accurate judgments, it is not always true and can lead to errors.

Consider, for example, the *gambler's fallacy.* If we flip a coin six times and the coin turns up heads on all six flips, many of us tend to think that on the next toss, it will turn up tails because "it's due," or the coin is rigged. Of course, this is incorrect. The coin has no memory, it has no way of knowing how long it's been since the last tail turned up. So, the likelihood of a tail turning up on any one toss is independent of what happened on the previous tosses. There is no way that the previous tosses can influence the next toss. The probability on the seventh toss is 0.50, just as it was on the first toss. Why does this very common judgment occur? We all know that with a fair coin, the chances of flipping a head or a tail is 50–50; therefore, we assume that what is true for one toss is also true for a sequence of tosses.

We incorrectly think that tossing a coin once is representative of all sequences of tosses. But this isn't true; some sequences of tosses are 75% heads; some are only 5% heads. It is only when we combine sequences over a large number of tosses that the probability moves closer to a 50–50 split.

The representative heuristic is also at work when we encounter the "man who argument." Imagine, for example, you are planning to buy a car of a certain make — Bomo make — because of the very good ratings this type of car gets from various consumer magazines. You mention your plan to a friend, who responds with shock: "You must be crazy. I know someone who bought a Bomo and the transmission fell out within the first month he owned it. Then the alternator went. Then the brakes went. Buying a Bomo would be a mistake." In this instance, your friend is offering a "man who" argument to persuade you. While consumer report magazines test many cars for their ratings, your friend presumably believes that his one example, "the man who" bought a Bomo, should persuade you not to buy this type of car. Your friend appears to believe that this one example is representative of the entire category or make of car. This "man who" argument is quite common and is very persuasive. Research has shown that people's judgments are often persuaded by one representative example rather than by studies with large samples of data.

Although the representative heuristic sounds like something to be avoided at all costs, it actually serves an adaptive purpose. Similar to the availability heuristic, at times it can help us make decisions (we have "gut" reactions for a reason!). However, it is important to remain nonjudgmental when making decisions and do the research for yourself (see Chapter 5). Simply being aware of how our brains can trick us can help us in making more informed decisions based on evidence instead of guessing or assuming.

Covariation

It is not surprising that we often use heuristics or mental shortcuts when we make judgments. It is, however, unsettling when we use these shortcuts when making important, consequential judgments.

The representative heuristic often shows up in arguments about climate change — for example, when someone uses the weather of a particular season to deny climate change over recent generations. Also, people use individual cases to support the link between vaccines and autism. At base, heuristics involve thinking about *covariation*. In brief, we can define covariation as when two events or attributes tend to occur in the same place or time: X and Y covary, if X tends to be on the scene whenever Y is; and if X is absent, then Y is usually absent. For example, exercise and stamina covary. Those of us, who do the first, tend to have the second. Owning a lot of CDs and going to concerts covary. Very often judgments about covariation are an attempt to understand cause and effect. And, it is true that there must be covariation in all cause and effect relationships. But we must be cautious. Not all covariation guarantees cause and effect. There are strong and weak covariations. The covariation between exercise and stamina is strong. The covariation between owning CDs and going to concerts is weaker; a good many people own many CDs and do not attend concerts. Also, while it is relatively easy to spot many positive covariations, where both attributes and events are present, it is not as easy to recognize negative covariation. While exercise and stamina positively covary — both are simultaneously present; exercise and body fat negatively covaries — a good deal of exercise is related to low amounts of body fat.

Thus, we often use thinking about covariation to make judgments about cause and effect. While it is true that covariation is necessary for cause and effect relationships, it is not sufficient to establish cause and effect. We must be cautious about drawing conclusions about cause and effect from instances of covariation. Does education lead to higher paying jobs? Does eating a good breakfast make you feel better throughout the day? Do aching joints indicate the approach of bad weather? Covariations can be strong or weak; they can also be positive or negative. And, while covariation must be present for any cause and effect relationship, not all covariation implies causation. There are many instances of *illusory covariations* — where two events or attributes co-occur suggesting a causal relationship when there is none.[24]

Illusions of causation derived from covariation are easy to document. Many people are convinced that there is a relationship between hand writing and personality, yet there is no research that documents such a relationship. Likewise, many people believe they can predict the weather by paying attention to arthritis pain, which turns out to be groundless. These illusions are not just bad or silly ideas. Those of us who have such ideas believe that we have had experiences of covariation that back up these beliefs. What causes illusions of covariation?

One explanation is based on the way we consider or process past experience as evidence for such beliefs. In most cases we seem to consider only a subset of evidence, and it's this subset that biases our judgments about our experience. This bias while essentially reasonable, guarantees mistakes. Specifically, when judging covariation, our selection of evidence is guided by the *confirmation bias* — that is, the tendency to be more responsive to evidence that confirms our beliefs rather than evidence that challenges or disconfirms our beliefs. This cognitive bias comes in many forms. First, when assessing a belief or judgment, we are more likely to notice evidence that confirms the belief rather than evidence that disconfirms it. We tend to have better memory for confirming evidence than for disconfirming evidence. When we encounter confirming evidence, we take it at face value; disconfirming evidence is often reinterpreted and distorted. When disconfirming evidence is made available, we often fail to use it to adjust our beliefs. And, finally we often fail to consider alterative hypotheses to explain our beliefs just as well as our first impressions.

The classic demonstration of the confirmation bias involves presenting participants with a series of numbers, such as "2, 4, 6."[25] The participants were told that the series of numbers conform to a specific rule and their task was to figure out what the rule is. The participants were allowed to propose their own series of numbers ("8, 10, 12") to test their discovery and the experimenter would respond appropriately ("Yes it follows the rule." or "No it does not follow the rule."). Then, once the participants felt sure of the rule, they announced their discovery. The rule was, in fact, quite simple: Any three numbers in ascending order. Thus "1, 2, 3" is correct; 31, 32, 33 is correct; "6, 4, 2" is not, and neither is "10, 10, 10." Despite this simplicity, participants had difficulty discovering the rule, requiring many minutes to complete the task. This was due to the type of information they requested to evaluate their proposal for the rule. To an overwhelming extent, they sought to confirm the rules they proposed; requests for disconfirmation were relatively rare. And, it should be noted that those few who did seek disconfirmation were more successful discovering the rule. It seems that confirmation bias was very much present in this study, and interfered with performance.

When we do encounter or seek out disconfirming evidence it is often reinterpreted so that we do not alter or readjust our beliefs. Gamblers who bet on professional football were asked about their strategies for picking teams and their record of wins and losses. These people all had "good" strategies for picking winners, and their faith in their strategies was undiminished even after a string of losses. Why? These gamblers didn't remember the losses as "losses." Instead, they remembered them as flukes or oddball occurrences. "I was right, and my team would have won, if it wasn't for that crazy injury to the running back." "Picking my team was correct, except for that goofy bounce the ball took after kickoff." Winning bets are remembered as "wins," but losses are remembered as "near wins." These gamblers retain their beliefs despite contrary evidence. This tendency is referred to as *belief perseverance*.

Examine the table below. This 2 × 2 matrix summarizes data from a study. The numbers represent the number of patients in each cell. Specifically, 200 patients were treated and showed improvement, 75 were treated and showed no improvement. Fifty received treatment showing improvement, 15 received treatment showing no improvement.

	Improved	No improvement
Treatment	200	75
No treatment	50	15

When asked to rate the effectiveness of the treatment, most people in this study thought the treatment was effective, and a number of individuals thought the treatment was substantially effective. In fact, the data in this table show the treatment is completely ineffective. In order to understand why, you need to concentrate on the outcome for the control group (the no treatment group). There we see 50 of 65 patients, or 76.9%, improved when they got no treatment. In contrast, 200 of 275 patients, or 72.7%, improved when they got treatment. Thus, the percentage of improvement actually is larger for the no treatment group. Why do we so easily make this mistake? When viewing the table we ignore the no treatment control group, which shows disconfirming evidence, and focus, instead, on the large numbers in the treatment/improvement cells, which seduces so many of us to confirm the treatment as effective even when it is not.

Reasoning

The confirmation bias is essentially illogical. A gambler should think: "If my gambling strategy is good, I'll win my next bet. I bet, but lose. Therefore, my strategy is bad." Instead, the gambler who has fallen prey to the gambler's fallacy thinks: "If my gambling strategy is good, I'll win my next bet. I bet, but lose. Therefore, my strategy is good." Historically humans have been touted as the "rational animal" capable of reasoning logically; yet, psychologists have long known that we humans are prey to error. Studies that examine how we solve even the simplest logic problems (syllogistic reasoning problems) regularly report error rates as high as 70%–90%. Much of the difficulty we have with logical reasoning has to do with beliefs and the language used to express a problem. Prior beliefs are ambiguous and abstract language can distort the logical form of arguments when reasoning. Some typical mistakes we make are as follows:

Affirming the Consequence (Converse)

(1) If A is true, then B is true.

(2) B is true.

(3) Therefore, A is true.

(1) If the object I'm holding is a frog, then it's green.

(2) The object is green.

(3) Therefore, it is a frog.

Denying the Antecedent (Inverse)

(1) If A is true, then B is true.	(1) If the object I'm holding is a frog, then it's green.
(2) A is not true.	(2) The object in my hand is not a frog.
(3) Therefore, B is not true.	(3) Therefore, it is not green.

We often conclude that these logical arguments are correct, but they are not. The examples on the left-hand side are given in abstract form and are quite difficult. On the right-hand side is the same argument written with a concrete example, and it is a little easier to see how they are logically incorrect. For affirming the consequence: If I'm holding a frog, it's green; the object I'm holding is green; therefore, it's a frog, does not follow. It could be a green pepper. Likewise, for denying the consequence: If the object I'm holding is a frog, then it's green; the object is not a frog; therefore, it is not green, does not logically follow. Again, it could be a green pepper or head of lettuce. In both cases, the conclusion does not logically follow from the two prior statements.

A great deal of research has examined our ability to reason about *conditional statements*: in an "*if x, then y*" format, with the first statement providing a condition under which the second statement is guaranteed to be true.[26] This type of reasoning is the basis for making scientific observations and drawing conclusions. Here too, typical error rates are as high as 80% or 90%. And also, *belief bias* plays a role: people will endorse a conclusion, if they happen to believe it to be true, even if the conclusion doesn't follow from the stated premises. Conversely, people will reject a conclusion if they happen to believe it to be false, even if the conclusion is logically demanded by the premise. Conditional reasoning is sometimes studied directly by having people evaluate statements but most often, the now famous *Wason selection task* is used because it nicely shows why we have problems with this kind of reasoning. In this task you are shown four cards with a number on one side and a letter on the other. The task is to evaluate the rule: "If a card has a vowel on one side, it must have an even number on the other side." Which card(s) must be turned over to put this rule to the test."

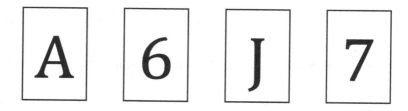

In the original research, 33% of the participants turned over the A-card to check for an even number. Another 46% turned over both the A-card and the 6-card. The correct selection was obtained by only 4% turning over the A-card and the 7-card. The A-card would have to be turned over: if there is an even number, this confirms the rule; if it is an odd number, this disconfirms the rule. The rule says nothing about a consonant, so it doesn't matter what we find on the other side of the J-card. Regarding the 6-card, the rule says nothing; a vowel or consonant can appear on the other side. Finally, the 7-card must be turned over: if there is a vowel this disconfirms the rule; if there is a consonant, the rule is confirmed.

Most people rightfully turn over the A-card to confirm the rule. But a good number of people also turned over the 6-card — attempting to affirm the consequent. The rule says nothing about what should be on the other side of an even number. And, only a few (4%) sought to challenge or disconfirm the rule by turning over the 7-card: if there's a vowel the rule is disconfirmed, if there's a consonant, then the rule is valid. This finding illustrates our inclination to seek confirming information and the difficulty we have challenging or seeking disconfirmation.

An interesting variation on this selection task problem shows that some of the difficulty we have with conditional reasoning has to do with having to deal with abstraction.[27] Participants were shown four cards and asked to test the rule: "If a person is drinking a beer, then the person must be at least 21 years old."

Drinking a Beer	22 Years Old	Drinking a Coke	16 Years Old

In contrast to very poor performance on the standard (A 6 J 7) version of the problem, performance on this version was much better. A total of 73% correctly turned the card labeled "Drinking a beer," confirming the rule, and they also turned over the card labeled "16 years of age," challenging the rule. They did not select the card labeled "Drinking a Coke" because anyone can have a Coke; and, they also did not select the card labeled "22 Years of Age" because they could have either a Coke or a beer. Thus, it would seem that how you think — and how well you think — depends on what you think about. The standard version and this modified version of the problem have the same logical structure but yield very different results. Having the script of "drinking" greatly facilitated seeing that logical structure.

Confirmation bias can cause problems in day-to-day life; for example, making inaccurate assumptions about others and consequently looking for those assumptions. Many people have had the thought, "That person doesn't like me!" and then only see information to support that thought. This can be an inconvenience when it happens. However, confirmation bias can become a serious problem in the fields of scientific research and the justice system, where many lives may be in the hands of others. There is a problem within the scientific community of publishing mostly articles with "statistically significant" results. However, the unpublished articles that have "insignificant" results are showing contradictory evidence that is often ignored, leading us to believe that many studies have "proved" the effectiveness of a certain drug or treatment, while selectively ignoring the studies that showed no result. Additionally, detectives and prosecutors may hone in on a suspect of a crime and fail to notice evidence that may point to another suspect or away from the current one. You can see how problematic this would be. How do we remedy this?

As mentioned several times previously, it is important to remain skeptical when presented with information and to never forget the importance of contradictory information. One piece of contradictory information can disprove a theory, while many confirming bits of information may serve as evidence but may never truly prove anything. Let's look at a simple example to display this. If you were asked to test the statement, "all rocks in the world are grey," you could find 10,000 grey rocks as evidence to support this. However, we all know that rocks can come in many colors, and all you would need to do is find one rock in any other color to disprove this statement.

Decisions

Choices, big and small, fill our lives, whether choosing courses to take next semester, which candidate to support in an election, or whether to marry. In many of these situations we don't have the luxury of making many observations or to experiment to evaluate possible outcomes. We must make decisions with limited information and no clear idea about which choice will get us closer to our desired goal. Traditionally, our decisions have been explained in terms of utility.[28] According to *utility theory*, when making decisions, we evaluate choices in terms of their potential *costs* (the extent to which the choice moves us away from our goal) and *benefits* (the extent to which the choice moves us toward our goal). In deciding, you weigh the costs against the benefits and seek the path that will minimize the former and maximize the latter. When you have several options, you choose the one that provides the best balance of costs and benefits.

Of course, this is not as easy as it sounds. Let's say you're looking for a warm vacation spot and you are considering either Tucson or Miami. Tucson's weather is better, but the flight to Miami is cheaper. To make a choice you need to weigh the pleasures of good weather against the savings in airfare. This contrast is hardly comparable. To make a choice you need to evaluate each factor in terms of your *subjective utility*; that is, ask yourself how important each factor is to you. In making a choice, you will need to calculate the utility of pleasant weather minus the "disutility" or cost of spending more money on airfare. And, beyond subjective utility, we need to factor in uncertainty: the weather in Tucson is not always great, while the weather in Miami is usually fine.

Consider a simple example: you are choosing courses for the next semester. One course looks interesting but has a heavy workload. First, you need to estimate the subjective utility of taking an interesting course against the cost of a heavy workload. Next, you need to estimate the chances that the course will be interesting and that it will have a heavy workload; let's say there is a 70% chance the course will be interesting, and a 90% chance the workload will be heavy. In this case, the overall utility of the course will be (0.70 × the benefit of an interesting course) minus (0.90 × the cost of a heavy workload). You would then make similar calculations for other courses and choose the one with the greatest overall utility.

This very rational approach to making decisions has been popular, especially among economists, throughout the twentieth century. However, as displayed through the heuristics and biases discussed in this chapter, psychologists have noted that very often our choices are influenced by factors other than a rational decision process. We will now discuss another aspect that biases our decision-making, *framing*.

Framing

When making decisions we are all powerfully influenced by factors having nothing to do with utility. It is relatively easy finding cases where this is so. The problems below are the classic examples that were used to establishing the *framing effect*:[29]

Problem 1

Imagine that the United States is preparing for the outbreak of an unusual disease, which is expected to kill 600 people. Two alternative programs to combat the disease have been proposed. Assume that the exact scientific estimates of the consequences are as follows:

If Program A is adopted, 200 people will be saved.

If Program B is adopted, there is a one-third probability that 600 people will be saved, and a two-third probability that no people will be saved.

Given this choice the majority of participants (72%) choose Program A, selecting the sure bet rather than the gamble. Now consider the rephrasing of the problem below:

Problem 2

Imagine that the United States is preparing for the outbreak of an unusual disease, which is expected to kill 600 people. Two alternative programs to combat the disease have been proposed. Assume that the exact scientific estimates of the consequences are as follows:

If program A is adopted, 400 people will die.

If Program B is adopted, there is a one-third probability that no one will die, and a two third probability that 600 people will die.

Here the majority of participants (78%) chose Program B, preferring the gamble rather than the sure bet. The puzzle, of course, is that the two problems are objectively identical: 200 people saved out of 600 is the same a 400 dead out of 600. Nonetheless, this change in how the problem is phrased — or how it is *framed* — turns a 3-to-1 preference in one direction into a 4-to-1 preference in the opposite direction. There is no right answer to either of the problems, either decision can be defended. The problem lies in the contradiction in choosing Program A in one context and Program B in the other context. This framing effect is so powerful that a single individual given both frames, on slightly different occasions, is quite likely to contradict himself. Another, less grim, example is presented below:

Problem 1

Assume you wish to be richer by $300 than you are today. You have to choose between:

   (A) a sure gain of $100,

   (B) 50% chance of gaining $200 and a 50% chance to gain nothing.

Problem 2

Assume you wish to be richer by $500 than you are today. You have to choose between:
 (A) a sure loss of $100,
 (B) 50% chance to lose nothing and 50% chance to lose $200.

When contemplating the first problem, almost three quarters of the participants (72%) chose option (A) — the sure gain of $100. When contemplating the second problem, the majority of participants (64%) selected option (B) — choosing to gamble. The two problems are identical. Both pose the question of whether you would rather end up with a certain $400 or with an even chance of getting between $300 and $500. Despite the equivalence, we tend to treat these problems very differently, preferring the sure thing in one case and the gamble in the second.

The pattern that emerges from these examples is that if the frame casts the choice in terms of losses, decision makers tend to be *risk seeking* — preferring a gamble, while hoping to avoid or reduce the loss. This preference is especially strong when people contemplate large losses. In contrast, if the frame casts the choice in terms of gains, decision makers tend to be *risk aversive* — they refuse to gamble, choosing to hold tight to what they have.

These examples illustrate how decisions can change depending on how *options* are framed. Similar effects can be found with changes in how a *question* is framed — should custody be "denied" or "awarded"? We are also influenced by how *evidence* is framed — for example, people are more likely to endorse a medical treatment with a 50% success rate than they are to endorse medicine with a 50% failure rate. Such effects make no sense from the perspective of utility theory. Framing should have no impact on the expected utility of these options. Yet, these differences in framing dramatically change the choices we make.

To keep from letting the way in which a problem is worded fool you, take emotion out of the equation. Ask yourself, "what is this actually saying?" It can be helpful in these cases to write down what is being asked in more general terms. If we had done this when looking at our gambling problem, we would have seen that we are choosing between identical options ($400 versus an equal chance of $300 or $500).

Emotions

Given that decision making is so fragile, it is no surprise that our decisions are powerfully influenced by emotions. For example, decision making is influenced by the emotion of regret. People are strongly motivated to avoid regret. Whenever possible, they select options that will minimize the chances for regret later on. Likewise, many decisions involve an element of risk, and the evaluation of risk affects the decisions we make. Specifically, we ask ourselves how much we would dread the experience when thinking about a nuclear accident or the side effects of trying a new experimental drug. Also, when considering a decision, remembering similar events can elicit bodily reactions — *somatic markers* — that can guide decision making. In such cases, we rely on our "gut feelings" to assess options, and this pulls us towards options associated with positive feelings and away from options that trigger negative feelings.

In addition to remembering past emotional events, predicting future emotional reactions can be an obstacle to good decision making. And, how good are we at making predictions about our feelings? Research shows that we are surprisingly poor. We tend to overestimate how much we will later regret errors. People give more weight to "regret avoidance" than they should. When asked to predict feelings about such events as breaking up with a romantic partner, learning you have a serious illness, failing to get a promotion, or getting a poor grade on an exam, people tend over estimate the amount of regret they might have. It seems we are convinced that things that bother us now will continue to bother us in the future. We tend to underestimate our ability to adapt.

While we all want choices in our lives, we are prone to errors when making decisions. In fact, having too many choices can make us unhappy — this is known as the *paradox of choice*. The difficulty stems from using strategies that leave us open to manipulation and self-contradiction. Our decisions are not strictly determined by the utility of the choices we are faced with. How choices are framed has a great impact on the decisions we make. And of course, errors in affective forecasting guarantee that people will often take steps to avoid regrets that, in reality, they might not have felt. Recognizing these limitations could help us make better decisions.

Summary and Conclusion

After a review of some of the difficulties in forming judgments, reasoning, and making decisions, some authors have concluded that humans are predictably irrational. We are, however, thought to be rational animals, uniquely designed to be capable of rational thought. There are many psychological aspects of thinking that can predictably distract us from rationality and lead us to error. When forming judgments, we cannot always take advantage of the richness of our experience. We rely on attribution substitution so that we estimate the frequency of experience by way of heuristics. Two common heuristics we often use to estimate the frequency of our experience are the availability heuristic and the representative heuristic. With the availability heuristic we estimate the frequency of an occurrence or attribute by the ease with which we can think of examples. With the representative heuristic we use a single case or occurrence as a basis for estimating a whole category of events. Both heuristics often lead to accurate judgments but they also can lead to errors. The source of such errors involves our misunderstanding of covariation and cause and effect relationships — which can lead to illusory correlations that are not causal. These errors can be explained as a bias in how we examine evidence. We tend to seek evidence that confirms our beliefs and have difficulty seeking evidence that challenge our beliefs. This confirmation bias is also a source of error in our reasoning. Generally, we are quite poor at logical reasoning but the errors in reasoning are not always the result of carelessness. They often derive from our tendency to confirm and maintain what we believe to be true, and disregard evidence that challenge our beliefs.

Chapter 4

Disinterest and Anxiety Versus Motivation and Confidence

As stated previously, there are several psychological factors that contribute to one's ability to effectively and efficiently solve problems. Most of the factors that were listed in the introduction (e.g., focused attention, planning) involve complex intellectual processes. It is possible to possess many of these intellectual factors already and still have trouble solving problems. You might wonder, what else could be getting in the way? That is why we will now discuss one of the most important factors that impede one's problem-solving abilities: anxiety.

What is Anxiety?

You are likely familiar with the term anxiety, which is a complex emotion that encompasses feelings of worry and nervousness and is often accompanied by physical symptoms, such as muscle tension, increases in heart rate and breathing, and restlessness. In some cases, a feeling of anxiety is adaptive. For example, anxiety about where one will go to college often helps motivate students to study hard to obtain the necessary grades. Additionally, fear of strangers can prevent children from talking to unknown adults and keep them safe from what might be an otherwise dangerous situation. However, there are many times when anxiety causes us to fear situations that are not dangerous. In these cases, anxiety often leads to negative beliefs regarding the feared situation and results in avoiding the situation. For example, if someone feels anxious about public speaking and is required to give a presentation at work or in school, he or she might have negative thoughts about how the presentation might go. The person might think, "What if I make a mistake or freeze?" "I might sound stupid," or "I don't know if I can do this." He or she might feel sweaty or dizzy and might have an elevated heart rate or hyperventilate. The person might have the urge to call in sick or ask someone else to give the presentation. All these factors encompass the feeling of anxiety, and these symptoms are often present when one feels anxiety in other situations, including working with mathematics (known as *math anxiety*) and solving everyday problems.

Many people struggle with math anxiety, which can affect their performance in the subject. Although math anxiety can stem from not understanding the material being presented, it can also arise because of the immense amount of pressure that people feel to be "good at math." If being "good at math," makes someone smart, then the opposite thought of failing at math is the assumption that I am unintelligent. That's a lot of pressure! People with math anxiety often panic when approached with a problem, and this increases their heart rate and breathing, and replaces effective, problem-solving thoughts with anxious thoughts. They might think, "I can't do this," "I'm so bad at this," "This is too much for me," or "I'll never get it." With all those anxious thoughts popping up, it's no wonder that people with math anxiety have trouble solving problems! They are often unable to concentrate on the problem and usually rely on others to provide step-by-step directions.

Individuals with math anxiety attempt to avoid math at all costs. Children and adolescents might skip school, college-aged students might avoid taking math classes, and adults might ask their friends to "split the check" at a restaurant or ask coworkers to complete basic calculations at work. These attempts to avoid math, help alleviate the anxiety in the moment; however, they are quick fixes and do not work to help someone with his or her anxiety in the long term. Additionally, avoidance of math can lead to missing out on learning necessary problem-solving skills which can decrease one's understanding of concepts and subsequently will later *increase* one's anxiety about math. While you might think that avoiding math will decrease your anxiety, it will most likely increase your anxiety in the long term. The quick fix solution to avoid math almost always backfires.

A quick note: Don't be fooled by those who appear to brag about being incompetent in math. As mentioned previously, a key component in anxiety is avoidance. If one uses being "bad at math" as a way of never having to do it, one avoids failing and thus avoid proving to others and oneself that he or she has failed. Although statements like this might be as simple as poking fun at oneself, oftentimes avoidance of math is also a significant factor.

It is important to work on reducing anxiety and replacing it with the confidence to become a better problem solver. By replacing math anxiety with confidence, other strategies in this book will come more easily to you. However, you might say to yourself that becoming confident is easier said than done. This chapter addresses that problem by exposing you to basic anxiety-reduction strategies that will increase your confidence about your abilities and decrease uncomfortable anxious thoughts and bodily sensations. Prepare yourself; as you will find out in the rest of this chapter, the best way to overcome your anxiety is to face your fears. Get ready to "walk through the fire" and exit on the other side with more confidence in your abilities!

In this section, we are going to teach strategies to increase your confidence in your abilities to do math and solve everyday problems. Confidence is more than an "I can do this" attitude. It is built from a number of factors: understanding the goals and variables of the problem, approaching problems with an objective and open mind, and feeling sufficiently prepared and capable to tackle the problem. Perhaps, the largest indicator of confidence is one's ability to cope with failure, which is often inevitable. This chapter will focus on how to prepare yourself to feel capable to solve problems and cope with failure. Our hope is that after reading this book, you will be equipped with all of the tools necessary to boost your confidence in your problem-solving abilities.

Let's take a look at the problem that could be a bit daunting at first read, but with a strategy of looking for a pattern, the problem becomes quite manageable. This sort of eye-opening event can certainly begin to build your confidence. The problem posed is to find the sum of the first 20 odd numbers.

By examining the problem, you will find that the 20th odd number is 39. Thus, we wish to find the sum of $1 + 3 + 5 + 7 + \cdots + 33 + 35 + 37 + 39$. Your first reaction might be to solve this problem by simply writing out all of the odd numbers from 1 through 39 and actually adding them. At best, however, this method is extremely cumbersome and time consuming, and there are numerous possibilities for making an error.

As mentioned in the previous chapters, when your experience with problem solving increases, so will your confidence to solve problems. Another possibility to grapple with this problem would follow the Gauss method mentioned earlier, which involves listing the 20 odd numbers as $1, 3, 5, 7, 9, \ldots, 33, 35, 37, 39$. Now, notice that the sum of the first and 20th number is $39 + 1 = 40$, the sum of the second and 19th number is also 40 $(37 + 3)$, and so on. This then requires determining how many 40's to add. Since there were 20 numbers under consideration, we have 10 pairs, and we multiply $10 \times 40 = 400$ to get the answer. Just experiencing this unexpected technique will take you one step further increasing your confidence.

We can also examine this problem by looking for another pattern, but in a different manner.

Addends	Number of Addends	Sum
1	1	1
$1 + 3$	2	4
$1 + 3 + 5$	3	9
$1 + 3 + 5 + 7$	4	16
$1 + 3 + 5 + 7 + 9$	5	25
$1 + 3 + 5 + 7 + 9 + 11$	6	36

The table reveals quite clearly that the sum of the first n odd numbers is n^2. Thus, the answer to our problem is simply $20^2 = 400$.

Alternative methods for solving a problem make future problem-solution efforts much more enhanced.

For example, here is a problem that can be overwhelming only because it appears confusing. The issue is to sort out confusion and put it into sensible terms.

If A apples cost D dollars, what is the cost in cents of B apples at the same rate?

There are several ways that you might tackle this problem. Most often, you might choose to use numbers in place of the letters, and then try to reinsert the letters to find the answer. This can easily lead to confusion, and, unfortunately, oftentimes to an incorrect answer. Some people have been taught to look for unit costs and proceed from there. Again, this too, may lead to confusion.

As a general rule, a problem like this one can best be solved by organizing the data in some meaningful manner. Here, we will use proportionality together with some common sense. The proportionality is obtained by setting up the proportion with the same units of measure in each fraction:

$$\frac{A}{B} = \frac{\text{cost of } A \text{ apples}}{\text{cost of } B \text{ apples}} = \frac{100D}{x}.$$

Notice that we have used common sense in obtaining the last fraction. Since the problem called for the answer in cents, we use a fraction with cents as the unit rather than dollars. Thus, when we find x, we have found the answer. The rest is simple algebra:

$$\frac{A}{B} = \frac{100D}{x}, \quad \text{and then } x = \frac{100BD}{A}.$$

While experience with these types of problems will certainly help us come to this simple solution more quickly, a lack of confidence often hinders us from stopping to think about how to frame a problem in a new way. Moreover, by solving problems "the long way," we trick our brains into thinking that these problems are more difficult than they really are. Let's look at ways to increase confidence so these problems do not occur as frequently.

Developing Your Interests

Confidence begins with taking an interest in a topic. Although some people find it easy to jump right into a new subject confidently (it is likely that these types of people do not have a fear of failure, which will be addressed soon), many people require a certain level of expertise in a topic before they are confident in their capabilities. By reading this book, you are already taking the first step in developing your interests!

Think about the things in your life that you find interesting — hobbies, activities, and topics. It is likely that the priorities and interests in your life have some purpose or meaning attached to them. Maybe you go for a jog every day because of the health benefits related to daily exercise, or perhaps you enjoy knitting because you can provide clothing and other handmade gifts to those you love. You could be a history buff and enjoy reading old books and educating yourself about the past because "history repeats itself." Individuals exhibit interest in activities that they find meaningful in some way.

When we find something "boring," it is partly because we do not see any meaning or purpose attached to it. It is possible that we have tried it and have not gotten the results we would have liked. If your problem is the latter, we hope that this book will assist you in building a foundation to become a strong problem solver, so that you can learn to solve both mathematics and everyday problems efficiently and accurately. If you struggle with the former, read on to help yourself find meaning in mathematics and problem solving.

To improve our math and problem-solving skills, it is important to first understand the importance and the purpose within our lives of expanding on our skills. As mentioned in the introduction, problem solving permeates every area of our lives. Think about your daily activities and the problems that you face each day. Which problems in your life were most difficult for you to solve? Wouldn't it be great if you felt confident in your capability to solve them? It can be helpful to make a list of the ways in which expanding upon your problem-solving skills would positively affect your life. Here is a sample list for your reference. Some of the points might apply to you.

How Increasing My Problem-Solving Skills will Affect My Life
- I will be able to solve problems faster and more accurately, which will save me time throughout my day.
- I will be more efficient and accurate in problem solving at work, which could lead to increased productivity and/or a promotion.
- I will feel more confident.
- I will be less reliant on others for help.
- I will feel more independent.
- I will be more confident taking on problems when I don't immediately know the answer.

Once you have written your list, try to imagine your life as a more effective and efficient problem solver. Think about the positive effect it would have on your life. Motivate yourself to improve your problem-solving abilities by keeping track of problems that appear in your life and the ease at which you can solve them. It can help to set a goal or deadline for when you would like to have improved your skill set. Once your goal is set, it is time to get working!

One way to take a proactive step in developing your interests is to watch others solve problems. You can read books about problem solving (like this one), watch online videos, and talk to people who are experienced in the area. Different people have varying perspectives on concepts and can share them at various levels of difficulty. More exposure to problem-solving strategies will result in an increase in confidence toward hearing and using the terminology and in your ability to practice problem solving on your own.

Lastly, the best way to develop an interest is to make it fun. Think of ways in which problem solving can be fun for you. Here is one that you might encounter in one form or another. Suppose you are organizing a basketball tournament but only have one gymnasium available for the games. Let's assume there are 25 teams competing in a single elimination tournament, where one loss means the team is out of competition. The question is how many games are needed to be played in the gymnasium in order to get one championship team. The typical method is to simulate a tournament by perhaps having 12 teams play against another 12 teams, and eliminating 12 teams in the first round. This process would be continued by counting the number of games played each round and then totaling all the games to determine how many are required to get a champion. Figure 4.1 provides a flow chart as to how one might count the number of games until you have a winner.

Although this careful method works, it would be much easier to consider looking at the problem from a different point of view. Let's consider how many losers there must be to get one champion amongst the 25 teams. Clearly, there would be 24 losers. Well, having 24 losers requires 24 games. There is your answer. You can see that sometimes looking at a problem from this point of view can be quite rewarding and time saving.

Apply math and problem solving to your areas of interest. Every hobby or activity has problems that must be solved at some point! In sports, different angles and velocities are more beneficial than others for kicking or throwing a ball, and teams often face challenges in working together to pass the ball effectively or steal it from the other team. In crafts and interior design, precise measurements and organizational skills are often vital. When you are able to apply problem solving to an area that you appreciate, problem solving will become more enjoyable and interesting for you.

For example, here is a problem that might arise if you want to redecorate your house by hanging pictures in frames on a large, blank wall. You aren't sure how large the wall is, and you have several sizes of pictures that need to be printed and framed. To further send you into a panic, you have a birthday party planned in five days, and you would like to show off the photos to family and friends. However, all you have done so far is select the photos. They aren't printed, and you don't have the frames. You aren't sure how to get this project done in time, or even if you have the skills to carry out the project to completion.

Someone who feels panicked by problem solving might opt to avoid solving this problem by deciding to forget the whole project or

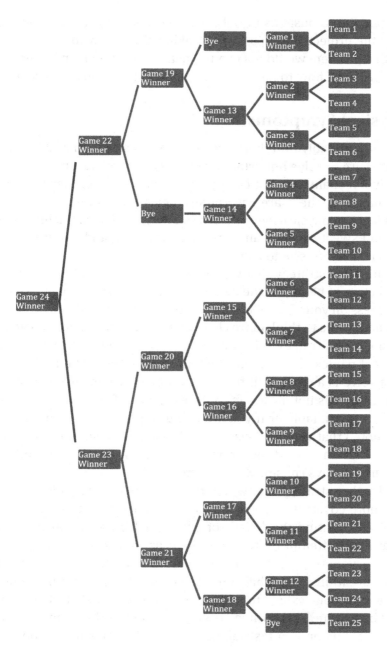

Figure 4.1

by spending thousands of dollars to have an interior designer complete the project. Perhaps, even thinking about how to go about solving this problem would make one's heart race or feel tense. Let's start by discussing how to handle those physical symptoms of anxiety.

Physical Symptoms

You will probably find that once you are in a situation in which there is pressure to solve important problems or there is a time constraint, you will feel anxiety that can interfere with your abilities to do so. If this happens, there are several strategies that you can use to help alleviate the physical symptoms of anxiety, which can include feeling hot and sweaty, having an increased heart rate and respiration, and experiencing muscle tension.

One such strategy that works to reduce heart rate and neutralize the feeling of being overheated is to splash cold water or hold a cold compress on your face. The cold compress or water will jolt your body into being able to think more clearly and will cool you down if you feel too hot. This strategy is best used when you are in a situation where you feel panicked. You might feel most panicked if the problem you must solve is a crisis or if there is time pressure involved. Of course, this strategy is not always feasible, especially if cold water or a compress is unavailable or if you are in a group setting. It should be further noted that very cold water can drastically lower one's heart rate, so use caution when deciding if this strategy is right for you, and when in doubt, always consult with a doctor.

Another strategy that works to reduce heart rate and respiration in a more covert manner is a technique called *paced breathing*. This involves counting the length of your inhales and exhales for each breath and attempting to make them as long as possible (without making yourself too uncomfortable). A common way to practice paced breathing is to inhale for 4 seconds and exhale for 6 seconds. It is ideal for your exhale to be longer than your inhale to help reduce your heart rate. This strategy is easiest to use in the moment of solving problems or in a test-taking situation because it is unobtrusive and not easily noticed by others.

A final technique to reduce physical symptoms of anxiety, particularly muscle tension, is progressive muscle relaxation (PMR). PMR involves tensing and relaxing different muscle groups in your body, usually in a systematic fashion. Practicing PMR can reduce stress and anxiety as well as increase awareness of tension throughout the body. This strategy is best used as a daily practice, either in the morning or before bedtime, or when you notice tension in your body. While it is often impractical to engage in PMR during a crisis or test-taking situation, PMR will help you reduce the amount of tension in your body and allow you to be in a calmer, more relaxed state.

As far as solving our home decorating problem, begin by practicing some strategies to reduce the physical symptoms of anxiety you might have.

- First, go into the bathroom or kitchen and splash cold water on your face. This will jolt your system, lower your heart rate, and counteract the feelings of being hot and sweaty.
- Take a few slow deep breaths. Count to four while you breathe in and then six while you breathe out. Do this for a full minute.
- Next, try to relax some of the muscles in your body. Relax your shoulders, arms, legs, and face muscles.
- Continue to breathe slowly until you feel your heart rate return to a normal state.

Some might argue, "How is this a problem-solving strategy? It isn't helping me solve the actual problem!" The truth is, physical symptoms of anxiety often *do* get in the way of solving problems, and reducing anxiety through strategies like relaxation and paced breathing can help you think more clearly and make more accurate and efficient decisions.

Cognitive Strategies

After tackling some of the physical symptoms of anxiety, it can be helpful to use cognitive strategies to further reduce your anxiety about a situation. Ask yourself the following questions when you feel anxious about solving a problem:

- "What is the worst that could happen in this situation?"
- "What is the best that could happen?"
- "What is most likely to happen?"

Think about the likelihood of the worst-case scenario occurring. Often, we believe that the probability of the worst happening is much larger than it actually is. Moreover, we usually overestimate the repercussions if the worst-case scenario actually does happen! Anxiety can be reduced considerably by taking a few minutes to examine the likelihood of each outcome and making predictions based more on reality and less on anxiety. When our physical symptoms become overwhelming, it can be harder to think clearly (and our predictions become much more inaccurate), so don't forget to use the strategies listed in the previous section before using these cognitive strategies.

Next, think about the problem-solving process and generate a plan both for the problem at hand and for how to proceed if your strategy is not successful. We often look at a problem and see the overwhelming possibilities of a negative end result, and this can cause us to quit or turn away. The same is true for a math problem. We see what we think is an overwhelming task and turn away. For both situations, it is helpful to break the problem into manageable steps. Additionally, visualize a good outcome to lower your stress. Picturing the situation going well will help you feel more confident about proceeding and will help prevent anxiety from interfering with the execution of your plan. If your strategy is unsuccessful, imagine coping with the failure in an effective way. Ask yourself, "What would need to change to increase the likelihood of a more successful outcome?"

Unfortunately, failure is an inevitable element of problem solving. Everyone fails from time to time; however, those with math anxiety live in constant fear of failure. What is it about failure that makes it so difficult for us to cope? The answer to this question will be different for everyone. For some, failure is linked to embarrassment or ridicule from others. For others, it is a desire to be perfect and an inability or unwillingness to recognize that imperfection in one area does not determine one's self-worth or abilities in other areas. It could be a fear of wasted time and effort, only to find out that the effort exerted was all for nothing. And for others there are likely additional fears associated with failure. It is important to ask yourself, when it comes to failure, what are you *really* afraid of?

Once you understand more about your fear of failure, challenge the thoughts you have about what it means to fail. Oftentimes, we "catastrophize" the results of failure (e.g., that everyone will laugh or think we are stupid, that all of our time was wasted with nothing gained). When the truth is that failure is a normal part of life and problem solving is often essential to learning. A good example is a batting average in baseball. A great batting average could be 0.300, and even a few Hall of Famers have an average below that number. However, a 0.300 means that the player only got 3 hits out of every 10 times at bat. That means the player failed far more than he succeeded, but people rarely focus on that point. Try to remember that example when fearing failure. Combat thoughts of not being successful at something by repeating to yourself that failure is a normal part of life and is to be expected from time to time. Tell yourself that it is better to have tried and failed than to not try at all. Remind yourself that *no one likes to fail*; however, it is inevitable and part of the process of overcoming obstacles in problem solving. The goal is to continually learn from mistakes, so you can improve and get things done. Telling yourself these positive statements to challenge your thoughts is called using *positive self-talk*. It can be most helpful to engage in positive self-talk before you even begin problem solving and then throughout the process to encourage yourself to persevere.

Let's apply these cognitive strategies to our redecorating conundrum and also look at them using mathematical formats. Start by asking yourself about the possible outcomes.

- **What is the worst that could happen in this situation?**
  I measure the wall incorrectly and purchase the wrong frames, which will lead to the wall being incomplete.

- **What is the best that could happen?**
  I complete all the correct calculations, purchase the correct frames, and complete the project that I proudly show to my friends and family.
- **What is most likely to happen?**
  I will probably make a mistake with measuring or purchasing frames, but I will get most of the project done by the party.

Next, generate a plan for solving the problem. This will help make the problem more well-defined. To begin, we might make a list of all the necessary materials that will be needed before starting the project.

Necessary Materials
- Tape measure
- Hammer
- Nails
- Pencil and paper
- Frames
- Printed pictures
- Large area (size of wall space) to lay out frames

Practice, Practice, Practice

In order for your interests to be further developed, one must practice extensively. Although practicing solving math and everyday problems will bring about anxiety in those who have difficulty, the best way to overcome anxiety is to face your fears. As you practice a task that doesn't come easy, you desensitize yourself to the triggers that activate anxiety. People with math anxiety often have similar fears about being unable to solve problems. The same strategies used to help with math anxiety also help with problem solving.

We all have skills in which we are naturally talented; however, think about something in which you worked hard to gain competence. Maybe it was a sport, a foreign language, or a musical instrument. At first, it may have seemed almost impossible without constant instructions or assistance from another person. However, over time, you developed your own skill set in the ability and were eventually able to practice on your own and feel more competent.

Math problem-solving and everyday problem-solving skills are developed in the same way. Because many believe that talent in math is a "natural ability," they assume that they cannot improve their skill sets and problem-solving abilities. While some have a natural ability in math (just as people have natural abilities in many different areas), it is not always the case that math either comes naturally or not at all! It is always possible to improve your math and problem-solving abilities. However, just like the acquisition of the other skills and abilities that you have learned throughout the years, acquiring skills in math and problem solving requires a lot of instruction and guidance at first. For example, learning multiplication for many children is a daunting task. They often need step-by-step instructions and repetition. However, now, you can likely multiply easily and quickly. Keep in mind that other mathematical and problem-solving skills are built in similar ways. They seem difficult at first, but with practice, you will gain competence.

The first way to practice solving problems is to start with more basic tasks. As mentioned in Chapter 2, most complex and difficult tasks can be broken down into smaller, simpler tasks that are easily completed. Therefore, before jumping right in to our redecorating problem, the smallest step we can take is to create a list of tasks that must be completed.

Tasks
- Measure wall
- Determine how large photos should be printed based on wall dimensions and number of photos being used
- Print photos
- Purchase frames
- Purchase (or gather) hammer, nails, and tape measure
- Place all photos in frames
- Lay out frames on open area the size of the wall dimensions and arrange until desired look is achieved

- Take measurements of space between photos and record
- Take photograph or sketch the arrangement to remember position of all photos
- Mark places on the wall where photos will be hung, based on photograph/sketch and space between photos
- Hang all framed photos

As you can see, even the simpler tasks have varying levels of difficulty. However, once the larger project is broken down into smaller, manageable pieces, it becomes less daunting and appears to be more easily achievable. You might then create a timeline to complete the project. In this example, let's pretend that you prefer to have the photos printed at a store, to ensure that the quality of the paper and image is excellent. You also already own hammer, nails, and tape measure. Below is the sample timeline.

Party Timeline

Monday	• Gather tape measure, hammer, and nails • Measure wall • Determine size of photos • Take electronically stored photos to store
Tuesday	• Purchase frames
Wednesday	• Pick up photos from store • Frame photos
Thursday	• Lay out frames on open space and arrange • Measure and record spaces between photos and take picture of arrangement
Friday	• Use pencil to mark the wall where photos will be hung • Hang all photos
Saturday	• Day of party! Enjoy your success!

Let's take a quick break from our party-planning project to discuss how there are quantitative aspects of all projects. Our current problem appears to be less about math and more about organizational skills and planning abilities. However, have you ever wondered how much space frame of a picture takes up? It probably would be an

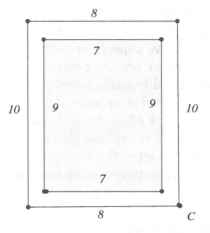

Figure 4.2

eye-opener to discover this, but more importantly it will make you more alert about the quantitative world around you. Let us consider a frame that has dimensions 8″ × 10″. A modest border of the frame might be ½ inch in width and not be considered obtrusive. Let us inspect that situation which is shown in Figure 4.2.

The area of the entire picture with frame is 80 sq. in., and the area of the photo is 63 sq. in. Therefore, the area of the border region is 80 − 63 = 17 sq. in. This happens to be $\frac{17}{80} = 0.2125 = 21.25\%$, or more than $\frac{1}{5}$ of the area of the page!

What is essential here is to be alert about the quantitative world around you. There are lots of examples in every-day life that could provoke this kind of astonishment. It is our hope that noticing this sort of eye-opening astonishment while completing a project makes it more fun!

Conclusion

In conclusion, building confidence in your problem-solving ability takes time, patience, and hard work. If you identify that anxiety is holding you back from facing math and problem solving, the best way to overcome it is to face your fears by practicing solving many different types of problems. Find ways to make math and problem solving fun by reading books that you find interesting and watching videos that portray problem solving in fun and creative ways. Notice the fun ways in which quantitative aspects of problems open your eyes to things you may have never noticed before. Physical strategies such as using cold water, paced breathing, and progressive muscle relaxation will help with physical symptoms of anxiety (e.g., racing heart, increased respiration) that often make situations more uncomfortable and lead to additional worry thoughts and catastrophizing.

Coping with failure is often the greatest hurdle that people face when they experience anxiety regarding math and problem solving. Cognitive strategies such as using positive self-talk along with perceiving failure as a normal aspect of problem solving can be useful in helping encourage you to persevere. Break down tasks into smaller, simpler pieces to transform daunting, complex tasks into ones that appear more easily achievable.

Chapter 5

Inattention and Forgetfulness Versus Focused Attention and Working Memory

Even when one is naturally skilled at problem solving, effectiveness and efficiency may be hindered if distractions are present or you forget key aspects needed to solve the problem. Inattention and forgetfulness are problems that appear to occur frequently in many individuals. Inattention is the inability to sustain attention for periods of time, while forgetfulness is the forgetting of already learned information.

Have you ever noticed that there are times you are able to concentrate for extended periods of time, while at other times you appear distracted by any stimuli that are in your presence? Or maybe you wonder why you can remember some (sometimes insignificant) details, but you have a more difficult time recalling information that you really want to remember? Although some individuals are naturally skilled at sustaining their attention for long periods of time and remembering key details, there are several actions that one can do to make it more likely that he or she stays focused on the problem. In this chapter, we will review several strategies to help build your ability to focus your attention and bolster your short-term memory.

Strategies to Help Focus Attention

Eliminate Distractions

It is not difficult to see how becoming distracted from challenging problems could cause delays in solving them. When we become distracted, we lose track of information that we have already established and make it difficult to continue a task fluidly. Experiencing continual distractions may cause you to have to restart the same task repeatedly because of difficulties remembering the problem goals, variables, and progress that you have made.

Examples of common distractions include: television, music with lyrics, cell phones, incoming email, and other people. Some people report working better with a lot of noise in the background, but the research shows that it generally is better to work without distraction. This will be discussed in greater detail later in the chapter. Turn off the television and music. Turn off the Wi-Fi if it isn't needed when working on a project on your computer. Leave your cell phone in another room or turn it off. Stop conversations with others (so long as they are unrelated to your problem, of course). A commonly overlooked distraction is room temperature. If your room is too cold or too hot, you might become distracted by the shift in your body temperature or your discomfort.

Additionally, the more difficult a task is, the more brainpower it requires and the more important it is that distractions are minimized. Think back to when you were first learning how to drive, a fairly-challenging task for most people. Many people report that before they were comfortable with driving, they had to drive with very faint or no music and could not hold a conversation while they were driving because of how strongly they had to focus on the road to drive safely. Even now, if you are an experienced driver, you might find yourself turning down the music or pausing conversations with others under unusual driving circumstances, such as driving in inclement weather or if you become lost. In these cases, our brains need even more energy to solve the problem at hand, which means minimizing or eliminating anything unnecessary.

Of course, this means that if you are solving easier problems, you do not need to allocate as much brainpower to the task as when you are solving harder problems. However, don't be fooled into thinking that distractions won't make a difference because you are solving easier problems. When the tasks are easier, efficiency and speed are key. Distractions can turn small tasks into large ones and easy problems into difficult and drawn-out problems. Consider the following problems that most people will find relatively easy. Complete them as quickly as you can with distractions present (e.g., television on, other people in the room talking), and record how long it takes you to complete the set.

$$
\begin{array}{ccccc}
24 & 76 & 43 & 16 & 47 \\
\times 5 & -9 & +8 & \times 4 & -8 \\
\hline
\end{array}
$$

$$
\begin{array}{ccccc}
364 & 568 & 44 & 281 & 26 \\
\times 4 & +23 & -36 & \times 9 & +3 \\
\hline
\end{array}
$$

Time to complete: _____

Next, complete the following similar set of problems with no distractions. Turn off your phone and any other electronic device that might distract you. Sit in a comfortable, quiet room. Time how long it takes you to complete the problems with no distractions.

$$
\begin{array}{ccccc}
27 & 72 & 36 & 41 & 97 \\
\times 4 & -7 & +9 & \times 2 & -6 \\
\hline
\end{array}
$$

$$
\begin{array}{ccccc}
234 & 962 & 14 & 108 & 33 \\
\times 8 & +15 & -1 & \times 6 & +8 \\
\hline
\end{array}
$$

Time to complete: _____

You likely noticed that the second problem set took less time to complete. Maybe you even made a few mistakes on the first problem set because you were not completely focused on the task. Although those sets only consisted of several problems, imagine if you had many more to complete. More problems would lead to more mistakes and longer time spent working on them. Although most people reading this book are likely hoping to learn how to tackle lengthy and challenging problems, we often overlook the importance of being able to solve simpler problems quickly and accurately. In life, we encounter many simple problems daily (in addition to the several challenging ones). Oftentimes, we spend more time working on easy tasks than is actually needed because of not focusing the way we should. We then do not do as thorough or accurate a job because we are distracted. Minimizing distractions, even when problems are simple, can increase our accuracy and efficiency.

Many people claim that some distractions actually help them focus instead of being distracting. For example, it is common to listen to music as one is completing work. There are a few reasons why this seems helpful rather than distracting. It is possible that the music is calming or soothing, which can alleviate anxiety. As discussed in Chapter 4, anxiety often hinders problem solving; therefore, using music as an anxiety-reduction technique may be helpful. However, music could also be simultaneously distracting, particularly if it has lyrics or is catchy ("I love this song!"). If you find music helpful, our recommendation is that you listen to instrumental music. The melody will be calming without being overly distracting and may hide other distracting sounds that are outside of your control (e.g., construction occurring outside or noise from others).

In sum, minimize distractions as best you can, whether completing tasks that are large, small, challenging, or easy! As shown in the activity of completing simple math problems, the practice of minimizing distractions will help you complete tasks more quickly and accurately. In the next section, we will demonstrate how to focus your attention once those external distractions are minimized. Here are two mathematics problems you might try that are still simple, but are a step up from the simple addition and subtraction problems from before. Try the first one with a distracting sound in the background and the second one without such distraction.

An army had 5 soldiers to every 6 soldiers of its opponent. During the battle, the first army lost 40,000 soldiers and the second army lost 6000 soldiers. The ratio of soldiers was then 2:3. How many soldiers are left in each army after the battle?[a]

Walking to school, Charles finds and takes 352 steps per minute. The measures each step to be 1.5 feet. What is Charles's speed and miles per hour?[b]

Practice Mindfulness: Stop Multitasking, and Focus on One Task at a Time

Many people have heard of mindfulness, which is the practice of focusing one's attention on one thing at a time and focusing on the present moment (not to be confused with meditation, which is a particular type of mindfulness). In order to practice mindfulness effectively, it can be helpful to first eliminate external distractions (as mentioned in the previous section), such as turning off the television or keeping your cell phone in another room. However, many find it extremely difficult to focus their attention, even when distractions are minimal. This is because internal distractions, such as thoughts and emotions, can also interfere with our ability to focus our attention, particularly on the present moment and the task at hand.

Have you ever noticed that as you were completing work, you had the thought, "I wonder what I'm having for dinner tonight," or maybe, "Oh, I just remembered that meeting that I will have at work tomorrow. It's going to be awful!" From there, you might have experienced feelings of anxiety or anticipation about what might happen next, and before you know it, you've spent 15 minutes planning a detailed dinner, or how you might discuss a particular issue at work, when your original goal was to complete a report or a presentation! This is an example of our thoughts distracting us from the present moment.

For these reasons, a key component to the practice of mindfulness is noticing when your attention is wandering. If you notice yourself getting distracted by external stimuli, thoughts of events that

---

[a] The answer is 36,000 soldiers and 54,000 soldiers, respectively.
[b] The answer is 6 miles per hour.

happened earlier in the day, or what is to come later, practice turning your focus back to your goal in the moment. The practice of turning your attention back repeatedly is key in helping one gain competence in mindfulness. This will seem very difficult at first, but it should become easier with practice! You might want to come up with a mantra to repeat to yourself to help you redirect your attention, such as "okay, back to work," or even simply, "focus!"

In addition to working on minimizing external distractions and redirecting your thoughts back to your goal, it is also important to eliminate multitasking. Oftentimes, it is not just one task that we have to complete but several tasks in a short time period or complex tasks that consist of many components. While many people believe they are expert multitaskers, the truth is that multitasking causes more mistakes and makes people perform tasks more slowly than when they focus their attention on one task (or part of a task) at a time.

If you have many items on your to-do list, it will be most effective to prioritize the list and tackle each item one at a time. Attempting to jump around from task to task will run you the risk of becoming distracted and not completing as much as you could in comparison to focusing on one task at a time. For example, your to-do list for part of your day might look something like this:

To Do Today
- Work on report
- Do laundry
- Do dishes
- Change cat litter

You might decide that it will be most efficient to throw in a load of laundry before beginning your report, which could take a few hours to write. However, once you begin your report, do your best to set aside as much uninterrupted time as you can. Of course, sometimes interruptions are inevitable. The phone rings, an unexpected visitor drops by, or construction work outside makes it difficult for you to concentrate. As mentioned previously, try to control for whatever you can before you begin working. Find a quiet room with few distractions. Keep your phone in a different room or turn it off.

However, many of us fall prey to avoidable distractions and interruptions. We might be working on one task and notice that another area of the room is dirty, and think, *oh, I'll just go take care of that quickly*. Or we remember something else that we wanted to do that day and quickly shift gears to focus on the newly remembered task. If you think of something else while working on an important project, make a quick note instead of interrupting work to attend to it immediately (unless, of course, it is an emergency). Provide yourself with a set time that you allocate for attending to a particular problem. It may help you remain focused knowing that you only have a prescribed period of time to work. Otherwise, you may put off starting to work by engaging in something like productive procrastination. This is when you complete a task that is important and has been on your "to do" list for quite some time, but should not be have been started in competition with the higher priority project.

Similarly, when working on a task or solving a problem, focus on one aspect of the problem before moving on to the next. Every time you shift gears to a new task, your brain and body will take some time to adjust to the shift. If this happens frequently, you will end up wasting a lot of time. For example, if you are working on a report that consists of multiple sections (e.g., past sales, new opportunities for business growth, ways to improve future sales), work on one section at a time without jumping around within the report. Do not switch from past sales to future sales, and then go to opportunities before resuming your work on past sales. Stay the course and give yourself a goal of how much you want to get done before switching to another area. Similarly, if you are doing housework and focused on cleaning, do not jump from room to room. Start in one room and focus on how to most effectively clean that room before switching to the next one. Once your brain is in a mindset, it will be more efficient to stick with the same task instead of disrupting your flow of work.

To effectively keep your brain in the same mindset and avoid switching gears too frequently, it can be useful to keep a list of what you hope to accomplish. This is another example of how breaking down larger tasks into small pieces can help you stay organized and feel more able to complete seemingly daunting tasks. Here is an example of a step-by-step list that you might keep while cleaning your house.

Cleaning Tasks
- Kitchen
  - Throw away all garbage
  - Put away any food or other items
  - Wipe down counters and tables
  - Sweep floor
  - Mop
- Living Room
  - Unclutter by removing any objects that do not belong in living room (e.g., coats, books, bags)
  - Wipe down tables
  - Dust furniture
  - Vacuum floor

To effectively complete the task, go through the steps one at a time, trying not to focus on the next step until you have completed the previous one. Here is an example of math problem that relies on this same strategy — namely, taking one step at a time. We are using a problem from high school geometry.

You are given circle O, where $\overline{AB}$ is perpendicular to $\overline{CD}$, as shown in Figure 5.1. You are asked to find the diameter of the circle in terms of a, b, c, and d.

Solution: It is natural to seek to use the Pythagorean theorem since perpendicularity was given. Similarity may also come into play. However, these approaches usually leave the diameter out of discussion. This can lead to frustration. Let's go through this problem step-by-step.

1. The first step in this problem is to consider the relationship between $\angle CEB$ and arcs $\overset{\frown}{CB}$ and $\overset{\frown}{AD}$. Once the relationship that

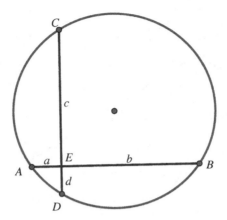

Figure 5.1

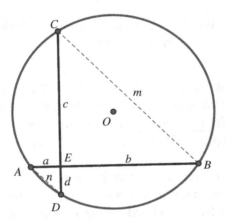

Figure 5.2

$m\angle CEB = \frac{1}{2}\left(m\widehat{CB} + m\widehat{AD}\right)$ is established, it is clear that $m\widehat{CB} + m\widehat{AD} = 180°$.

2. Next consider the two right triangles $\triangle CEB$ and $\triangle AED$, shown in Figure 5.2, and apply the Pythagorean theorem to each one. Then $m^2 = c^2 + b^2$ or $m = \sqrt{c^2 + b^2}$, and $n^2 = a^2 + d^2$ or $n = \sqrt{a^2 + d^2}$.

3. Let us do something a bit unusual now and create a simpler analogous problem (without loss of generality) that will help us solve

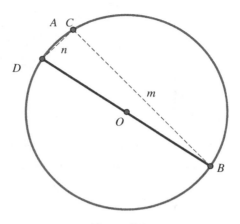

Figure 5.3

our problem. We will move arcs $\overparen{CB}$ and $\overparen{AD}$ along the circle so that they share a common endpoint, A and C, as we can see in Figure 5.3.

4. Now we can see that since $m\overparen{CB} + m\overparen{AD} = 180°$, $\overline{DB}$ must be a diameter of circle O, and since $\angle DAB$ is inscribed in a semicircle, $\angle DAB$ is a right angle. Therefore, diameter $BD = \sqrt{m^2 + n^2}$, which from the above equals $\sqrt{(c^2 + b^2) + (a^2 + d^2)} = \sqrt{a^2 + b^2 + c^2 + d^2}$.

Therefore, a simply stated problem has a surprise solution, one critically dependent on angle measurement with a circle, specifically, of the angle formed by two chords intersecting in a circle, which we noted is one-half the sum of the intercepted arcs. It is also important to see that we were able to move the chords to a more desirable position without any loss of generality and we then created a more easily solved problem. More importantly, you will notice that we needed to systematically take one step at a time to get to the next step.

Evaluate Nonjudgmentally — Stick to the Facts of the Problem

Another component of the practice of mindfullness is to gather information and evaluate that information in a nonjudgmental manner. At first, this can seem unrelated to problem solving. After all, what do our opinions and judgments about problems and solutions have to do with solving a problem? It turns out that opinions and judgments can cloud our thoughts and hinder our ability to problem solve. Additionally, we can become distracted from the goals and variables of the problem when we impose our own opinions on it.

First, we will explain what it means to be judgmental. When we use the word *judgmental*, we really mean using language that places a value or imposes an interpretation on something. For example, calling an idea "good" or "bad" is a judgment because others may disagree with that interpretation. Similarly, calling a problem "hard to solve" or "unfair" or saying that "it doesn't make sense" are also judgments. More self-explanatory judgments are calling oneself or others "stupid," "mean," or any other choice word that one might use out of anger. This is a problem because judgments often evoke negative feelings such as anger and anxiety, which can drastically interfere with one's problem-solving abilities (see Chapter 4). Becoming over emotional about a task almost never helps with focusing and usually serves the opposite purpose. Judgments rarely give us a next logical step for how to go about actually solving the problem. For example, saying that a problem "doesn't make sense" will likely make you feel angry and anxious and will create a wall between you and the problem, making it less likely that you will want to tackle it. Thinking that a teacher, professor, or boss is "mean," or "unfair" also increases one's level of anger and does not describe the problem in a solvable way. Remember, this book is about being able to use psychological strategies to *solve problems*. Usually, when we use judgmental language, we aren't solving problems. We are complaining.

So, what should you say instead? It can be challenging to reword statements nonjudgmentally; however, doing so can be incredibly beneficial. In contrast to judging, practicing using nonjudgmental language can lessen negative emotions such as anger and anxiety. In place of saying that a problem "doesn't make sense," try defining the problem in a way that sticks to the facts and describes where you are facing difficulty. Saying instead, "I don't understand what I need to do to solve this problem or where I should begin," provides you with the next logical next step — finding someone who can explain it so that you understand what to do. Instead of saying that a professor is "mean and unfair," take ownership of your own thoughts and feelings about the situation by saying, "I notice that I feel angry because the professor is unresponsive to emails and gives pop quizzes." Now you have a way to actually generate solutions! You could attend office hours to ask questions in person and develop a study plan to manage the pop quizzes. If it's the professor's demeanor that you have issue with, then by all means develop a support group of students to complain to each other. Keep in mind it won't solve the problem of the emails or quizzes (or the professor's demeanor, for that matter), but you might all feel a little more cohesive and understood in the process!

To help you get started, here are some common judgmental statements with examples of ways to phrase them using nonjudgmental language. When formulating a nonjudgmental statement, keep in mind that the point is to stick to the facts to help you reach a solution or to help you generate a plan to get started.

Judgmental statement	Rephrased nonjudgmental statement
That person is (insert negative adjective here).	I dislike that person because... (describe person's actions).
This problem is too hard.	I am not sure what I need to do to solve this problem.
This problem is harder than the rest of the ones in the set.	I'm noticing that there are more components to this problem than there were in the other problems.
I'm so stupid.	I feel upset/annoyed/angry with myself for my incorrect answer.
I'll never get it done.	I feel overwhelmed because I think this is going to take longer than I expected.
This is ridiculous/pointless!	I don't understand the purpose of this.
I'm never going to get it.	I feel frustrated at myself for trying and failing repeatedly.
This is so annoying.	I feel annoyed about having to complete this work.

In the following example, this problem could appear very frustrating but if looked at in a step-by-step fashion it can be solved without too much trouble.

> To win an election, Max needs 3/4 of the votes cast. If after 2/3 of the votes have been counted, Max has 5/6 of what he needs to win. What part on the remaining votes does he still need to win the election?

Notice the thoughts that you are having regarding solving the problem. At times, you may have thoughts that the statement of problems can be confusing and frustrating. However, practice rephrasing any judgments that you have to statements that can assist you in tackling the problem. Instead of thinking "this problem is too confusing or complicated," tell yourself "I can solve this more easily if taken step-by-step."

Let V be the number of votes needed to win. The statement of the problem can be confusing and frustrating. However, it can be easily be solved by using a step-by-step process. After $\frac{2}{3}$ of the votes, which we will represent as $\frac{2}{3}V$ the office, have been counted, Max has $\frac{5}{6}$ of what he needs to win, which is $\frac{3}{4}V$. Therefore, he now has $\frac{5}{6}\left(\frac{3}{4}V\right)$ or $\frac{5}{8}V$. Since he needs $\frac{3}{4}V$ to win, he still needs $\frac{3}{4}V - \frac{5}{8}V = \frac{1}{8}V$. However, $\frac{1}{3}$ of the votes have not yet been counted. Hence, $\frac{1}{8}V / \frac{1}{3}V = \frac{3}{8}V$ represents the part of the remaining votes that Max needs to win the election. Patience and a step-by-step approach wins the day.

Strategies to Increase Working Memory

In addition to being able to focus one's attention, many people wish to increase their memory capacity. However, increasing one's capacity for remembering information is complex, as memory is comprised of three subtypes: sensory memory, working memory (also called short term memory), and long-term memory. Sensory memory only lasts a maximum of few seconds and consists of information taken in from our senses, such as hearing a sound or seeing words on a piece of paper. We do not attend to much of the information that enters our sensory memory. However, when we attend to information from our sensory memory, it enters our working memory, which often has a span of a several seconds to a minute. We can store about 7 (±2) pieces of information in our working memory. If we continue to attend to information in our working memory, it may be consolidated into long-term memory, which has an undefined capacity and can last an indefinite amount of time, assuming the information is attended to occasionally. When information leaves any of the memory stores, it is known as *forgetting*. It is essential to bolster one's working memory because it plays an important role in problem solving. If information is misremembered or forgotten, it may cause errors or be left out of the problem altogether.

When people complain about having poor memory or continually forgetting information, they are most often referring to the consolidation of information between working memory and long-term memory. For example, you may have placed your keys somewhere new and are unable to remember exactly where the keys are located several hours later. However, there are several reasons for why you may be "forgetting" information.

One reason that people are unable to remember information is that they did not actually properly attend to the stimuli in the first place! If this is the case, the information never entered sensory memory to begin with, or the information did enter sensory memory, but it was unattended to and did not have the chance to move to working memory. This often happens when people are distracted by other external or internal stimuli. Examples of external stimuli (as mentioned in the previous section) could be distractions such as the television on in the background or texting while having a conversation. Do not expect to remember much of a conversation that you are having while you are also playing the latest popular game app, or texting a friend on your mobile phone. Internal stimuli include thoughts and body sensations. Sometimes we may appear to be paying attention to a conversation or lecture, but our brains are preoccupied by other thoughts (e.g., *When is lunch?* Or, *I have so much work to do later...*). In sum, if we are not paying attention to the information in the first place, it is unlikely to be remembered at a later time. Refer to the previous section for strategies on how to stay focused and attend to information that is vital to solving problems.

Remember More

Another reason that people often forget information is that they attempt to hold too much information in working memory. Remember, working memory capacity is only 7 (±2) pieces of information and lasts up to a minute. If you attempt to remember a 15-digit number, you will have much more trouble keeping that information in working memory than you will in remembering and recalling a 6-digit number. It can be frustrating to complete long, drawn out problems, with such a limited working memory capacity; however, it is possible to trick your brain into remembering more than 7 (±2) pieces of information.

One way to trick your brain into remembering information is to chunk information into more memorable pieces. It is likely that at times, you do this without thinking. The most common way that people chunk information to remember is with phone numbers. It would be difficult to remember the phone number 8-2-1-5-5-5-6-2-9-7 (10 digits) if you were simply remembering each digit in succession. However, most of the time, the phone number is broken into chunks: (821)-555-6297. Because area codes (e.g., the 821 in this example) often have meaning, they are usually easy to remember as one "chunk" of information. The next numbers, however, may be more difficult to remember. Therefore, you may decide to further chunk the last 4 digits into two, more-easily remembered, numbers, 62 and 97. Now, you have to remember (821)-5-5-5-62-97, which are 6 chunks of information.

A specific form of chunking is making use of acronyms and mnemonics. Acronyms help us to remember information by using the initial letters of key words to make another, more easily remembered, word. Mnemonics use a pattern of letters, ideas or associations that lead to an easier retrieval of information. These strategies are incredibly useful in helping one learn large amounts of information easily.

For example, students learning neurology must learn the twelve cranial nerves in order. In case you have an interest in the area, here are the names of the nerves:

1. olfactory

2. optic

3. oculomotor

4. trochlear

5. trigeminal

6. abducens

7. facial

8. auditory (or vestibulocochlear)

9. glossopharyngeal

10. vagus

11. spinal accessory

12. hypoglossal

Not only are those names complex and difficult to pronounce, but also remembering them in order makes things even more challenging. Some clever students created the mnemonic: *Ooh, ooh, ooh, to touch and feel very good velvet. Such heaven!* While this mnemonic doesn't help anyone recall the complex names (or spelling, for that matter) of the cranial nerves, it uses the first letter of each name of a nerve and creates a sentence that one can easily remember and refer to while taking an exam.

There are a number of mnemonics and acronyms in mathematics that students use to remember certain processes. For example, FOIL is used to recall how to multiply two binomial expressions. **FOIL** represents **First–Outer–Inner–Last** for multiplying $(a + b)(c + d) = ac + ad + bc + bd.$

Here are a few more such acronyms that help people remember problem-solving procedures:

To remember the order of doing arithmetic presented as a series of operations, some use: **BIDMAS** — **B**rackets, **I**ndices, **D**ivide, **M**ultiply, **A**dd, **S**ubtract.

To member the definitions of the trigonometric functions some use: **SohCahToa** to compute the sine, cosine, and tangent of an angle. **Soh** stands for **S**ine equals **O**pposite over **H**ypotenuse. **Cah** stands for **C**osine equals **A**djacent over **H**ypotenuse. **Toa** stands for **T**angent equals **O**pposite over **A**djacent.

Write Down Information

A third reason that people often forget is because they wait too long between attending to the information and then accessing the information later. If you want to remember to bring your leftovers to work the next morning, thinking about it the night before might not do you much good. Simply, too much time has passed, and the information has left your working memory. This may also happen during long word problems. By the time you have finished reading the problem, you might barely remember the information presented in the beginning of the problem. The information entered your working memory, but then more information came in and too much time passed for you to process and remember all of it.

For this reason, it can help, and sometimes is necessary, to write down useful information. This will ensure that relevant information is not forgotten. For example, consider the following word problem, in which you may find it helpful to write down information to remember what the problem is asking.

Here is one example where it is helpful to write down the material step-by-step:

A fish has a head that is 7 inches long. Its tail is as long as its head, and one half as long as its body. The length of its body equals the length of his head and tail together. What is the length of the entire fish?

The solution requires us to note each of the pieces that we were given one at a time as follows:

> We will let t be the length of the fish's tail, and b be the length of the fish's body.
>
> We get that the length of the fish's tail, $t = 7 + \frac{1}{2}b$, and $b = 7 + t$. By substituting for b in the first equation, we get $t = 7 + 2(7 + t)$, which is $t = 21$. It follows that $b = 28$. Therefore, the length of the entire fish is $h + b + t = 7 + 28 + 21 = 56$.

We can apply this strategy of writing information down to everyday life. To help yourself remember to bring an item with you to work or to complete a task before you leave the house, use sticky notes and place them in high-profile areas. For example, stick a note on the refrigerator to remind you to bring your leftovers to work. You will see it when you open the fridge to have breakfast that morning and remember to take the leftovers with you. Or leave a note on the door that you exit each morning (e.g., reading, *Did you feed the fish?*). A similar strategy that does not require writing is to leave items that you wish to remember with high-priority items that you cannot forget, such as your car keys. It is impossible to leave the house without your car keys; therefore, you will also remember the item placed beside them. These reminders place previously forgotten information back into your working memory for just the right amount of time. Once you are finished with them, they will be forgotten and new information can take priority.

For remembering day-to-day information, such as appointments and events, it is extremely useful to keep a planner. Write down information that you receive immediately to avoid having it leave your working memory. Keep the planner with you at all times so that you do not forget to write something down at a later time. In this era of technology, many people find the calendar or notes apps on their phones to be very useful to record this kind of information. However, once the calendar is out of sight, the information might be forgotten. That is why another strategy that can work to bolster your memory is to keep a desk or wall calendar in your home or office. Each time you sit at your desk or walk by, you will be reminded of upcoming events, keeping the information consistently in your working memory.

Understand the Information

Lastly, occasionally people do not really understand the information that they are attempting to remember, which makes it more likely that they will forget. Many people can relate to this. If you are in school, you might notice that when you understand the information being taught, you are more likely to remember it. However, when you do not understand what your teacher is talking about, you will have a much harder time remembering later what was said. It isn't that you weren't paying attention! When people do not understand information, they are likely holding or replaying the information in their working memory stores but not consolidating it to long-term memory. Therefore, after a few minutes, much of the information is forgotten. It makes sense, then, that the more you understand, the better you will remember. While there are numerous ways to "better understand" material, there are a few that span across disciplines, which we will review here.

One way to better understand information is to *stop* doing something — that is, making flashcards. While many people find flashcards to be helpful study devices, oftentimes they fuel a mentality of memorization. You should NOT attempt to memorize anything that you do not understand, especially if it is a topic in which you will be tested later. It is very likely that the question might be worded differently, or that you may forget parts of the definition that you memorized. While there is some merit to memorization, the information will be much better remembered, if you make some meaning out of it. Next, we will discuss some ways to do just that.

Instead of making flash cards or writing and rewriting notes, consider making a diagram of information you want to remember. This could be a timeline of dates with important events, a drawing of the human body or a cell with labeled parts, or an information tree that breaks information down into smaller parts. Diagrams work for several reasons. First, the information is being rehearsed, which makes it more likely to enter long-term memory. Next, the material is being categorized into chunks of information, which aids in memory (as described previously). Finally, the information is being structured in a meaningful way that makes sense to the individual organizing it.

A final, arguably more enjoyable way, to remember and make meaning of information presented is to present or organize the information in a fun format. You might remember back in grade school learning a song to help you remember the names of all fifty states in the United States of America. More recently, there are songs to help one remember all the elements on the periodic table of elements, and a story to help remember the US presidents.

One way to assess how well you understand information, besides, of course, taking an exam, is if you can teach the material to someone else. Being able to teach the material shows a level of understanding information that is more thorough than being able to answer questions on a test. Try to explain the information to someone that does not know any of the jargon or technical terms in the subject you're trying to teach. If you can get them to understand the concepts, then you know you have just about mastered the material.

Let's try a word problem that requires the use of all the memory skills we have discussed. To solve the problem, you will need to simplify the information into a diagram and/or more manageable pieces (chunking), write it down, and understand the goal of the problem.

Consider the problem:

> *Among 40 Girl Scouts in one division at Camp Ellwood, 14 fell into the lake, 13 came down with poison ivy, and 16 were lost on the orientation hike. Three of these had poison ivy and also fell into the lake. Five of them fell into the lake and also got lost. Eight came down with poison ivy and also were lost. Two of them experienced all 3 mishaps. How many of the Girl Scouts in this division escaped with none of these mishaps?*

Now consider the solution, noting all the given information and arranging it in a logical fashion. Traditionally, most begin to solve this problem by adding all the cases given, and then subtracting duplicate things that happened to people. Rarely is this procedure effective.

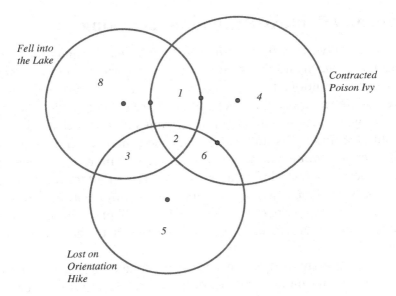

Figure 5.4

Let's examine the problem with a visual representation. As we have in Figure 5.4, we make a drawing to show the data with a set of circles (a Venn diagram):

The area of overlap for all three circles contains the 2 Girl Scouts who were lost, fell into the lake, and contacted poison ivy. The circles reveal:

Lake	= 14	Poison ivy + lost	= 8
Lake + poison ivy	= 3	Poison ivy	= 13
Lake + lost	= 5	Lost	= 16

$$\text{Total} = 8 + 3 + 2 + 1 + 4 + 6 + 5 = 29$$

There were $40 - 29 = 11$ who suffered none of these mishaps.

State and Context-Dependent Learning

A final strategy that one can call upon to more easily remember information is to learn and recall information in the same emotional state. You may have noticed that when you are sad or angry, you tend to have a better memory for other sad or angry events, while when you are happy, you can easily recall other times you experienced happy events. Of course, it is not always possible (and sometimes impossible) to choose your emotional state in a given moment; however, it can be helpful to employ strategies to create an effective emotional state in which you will learn and recall information. Consider using the anxiety-reduction strategies from Chapter 4 to help create a calmer and more focused mood from which to learn and recall information.

Similarly, many people have an easier time recalling information in a similar environment or location to which the information was learned. For example, you have a better chance of recalling information on an exam if you previously learned that information in the same classroom (or a similar room) in which you are taking the test. This is because the cues in the environment are similar during learning and recall, which help create associations in your brain and jog your memory when you need to retrieve the information. Considering the information on mood-dependent and context-dependent memory, maybe your boss has a point when he thinks you are less efficient when you work from home. Not only is your mood likely different (odds are, you are more relaxed); but the environmental cues are also totally different, potentially throwing off your usual flow of work!

Conclusion

In conclusion, focused attention and strong working memory are vital skills in problem solving. When we stay focused on a task, we are able to solve problems more accurately and efficiently. Likewise, the ability to hold information in working memory is crucial in ensuring that no information from a problem is forgotten. However, many people struggle with both carefully attending to a problem and remembering all of the necessary details of it. Because of this, it is important to learn strategies to promote focused attention and bolster working memory.

Focused attention is the ability to sustain attention for extended periods of time. It is more likely that one can attend to information by first minimizing distractions in the environment, including electronic devices and conversations with others. Next, practicing mindfulness and eliminating multitasking help to ensure that the information is continually attended to. Mindfulness can help one redirect one's attention continually to stay focused on one task at a time before moving on to the next task. Finally, practicing focusing on one's goal to solve the problem in a nonjudgmental manner that sticks to the facts of the problem is essential in gaining clarity for how to solve it. Judgments often cloud our decision-making process and can distract us from making an effective decision that moves us toward a solution to the problem.

Working memory is the ability to recall information that has been attended into and entered our short-term memory store. While many people complain about problems with their short-term memories, sometimes the information that they "forgot" was never attended to in the first place! Therefore, it is important to ensure that one is using the strategies to increase focused attention in addition to bolstering working memory. Strategies that can help aid in memory of information include chunking the information into pieces and creating mnemonics or acronyms, which is another form of simplifying information. It is also important to write down important information to ensure that it is not forgotten, especially if it is expected that large quantities of information is to be remembered. When learning new material, it is essential that one understands what is being taught so as to ensure that it is better encoded into long-term memory, as opposed to being held in working memory as memorized material. A final memory strategy is to learn and recall information in the same mood and environment, increasing the likelihood that internal and external cues are similar to create associations between what is being learned and the mood or location you are in. Keep in mind that while attention and memory strategies can be beneficial tools to add to your arsenal of problem-solving techniques, they, like the other strategies in this book, are meant to be used in conjunction with the other techniques reviewed.

Chapter 6

Thinking Forward and Backward: Intuitive and Deliberative Thought

Most of us readily accept the old proverb that the best way to acquire a skill or get new information is through experience: "Experience is the best teacher." If you want to be good at problem solving, work on as many varied problems as you can. Practice, practice, practice! Yet, it is not that simple. Repeated unsuccessful experiences can add to frustration. Anticipating failure adds anxiety. And, the rote repetition of a solution hardly amounts to knowledge or understanding. Some sort of guidance is needed to benefit from experience.

The great Argentine writer Jorge Luis Borges wrote a wonderful short story, *Funes the Memorious*, about a young man living in Paraguay, who was kicked in the head by a horse and acquired the remarkable capacity for remembering specific experiences[30]:

> He knew by heart the forms of the southern clouds at dawn on the 30th of April, 1882, and he could compare them in his memory with the mottled streaks on a book in Spanish binding that he had seen only once ... These memories were not simple ones; each visual image was linked to muscular sensations, thermal sensations, etc. that could reconstruct all of his dreams ... Two and three times he reconstructs the whole day; he never hesitated, but each reconstruction had required a whole day.

Borges's story was until recently considered a fantasy. But in 2006, researchers published a case study of a patient they referred to as AJ. A lot like Funes, AJ remembered just about everything she experienced, every tiny detail of every meal she's ever eaten and every social interaction she ever had. She explained:

> I am thirty-four years old and since I was eleven, I have had this unbelievable ability to recall my past, but not just in recollections... I can take dates between 1974 and today, and tell you what day it falls on, what I was doing that day and if anything of great importance ... Whenever I see a date flash on the television (or anywhere else for that matter) I automatically go back to that day and remember where I was, what I was doing, what day it fell on and on and on and on.

This condition is called *hyperthymesia*, or highly superior autobiographical memory. It is exceedingly rare; seen only in a handful of people. AJ's ability seems miraculous and somewhat comparable to the capacity of a computer — where a thumb drive, the size of a pack of gum, can hold the equivalent of almost two million copies of text, 200 songs, and 300,000 photographs. But, if remembering past experiences is so important, why is hyperthymesia so rare? Why don't we all have similar abilities? Research indicates that for most of us memory of past experiences is quite poor and subject to distortions. An answer is that our mind was not designed to record experience in exact detail. Our mind, designed by evolution to solve specific kinds of problems and remembering tons of details, doesn't help achieve those solutions. Borges knew this when he noted Funes's comment: "I alone have more memories than all of mankind has had since the world has been the world ... my dreams are like you people's waking hours ... My memory, sir, is like a garbage heap." AJ also describes her experience of hyperthymesia as a terrible experience:

> It is nonstop, uncontrollable and totally exhausting. Some people call me the human calendar while others run out of the room in complete fear but one reaction I get from everyone who finds out about this 'gift' is total amazement. Then they start throwing dates

at me to try and stump me ... I haven't been stumped yet. Most have called it a gift but I call it a burden. I run my entire life through my head every day and it drives me crazy!!!

AJ is not alone in struggling with her condition. A story on NPR, 2013, reported that fifty-five hyperthymesics have been identified, and most struggle with depression.

Another example of how our mind works is in solving the problem of face recognition.[31] Humans are tremendously skilled at face recognition, despite this being a particularly hard information processing problem. Back in 1966 it was thought that we possess a "grandmother cell," a hypothetical neuron that represents a complex but specific concept or object. It activates when a person sees, hears, or otherwise sensibly discriminates a specific entity, such as his or her grandmother's face. However, most of the reported face-selective cells are not grandmother cells since they do not represent a specific percept, that is, they are not cells narrowly selective in their activation for one face and only one face regardless of transformations of size, orientation, and color. Even the most selective face cells usually also discharge, if more weakly, to a variety of individual faces. Yet, people are capable of discriminating between thousands of slightly different faces, and moreover, we need to recognize the same face under many different conditions. Every time we see a face, it is at a different orientation in our visual field, with slight differences in lighting, location, makeup, or shadows. If we tried to recognize faces based on an exact sensory experience of a specific neuron, we would fail miserably. We must pick out deep properties of a face that are present in every view we have of the face rather than recording the exact image of the face. This allows us to distinguish one person's face from another. The relative positions of different features are important aspects in face perception. It seems we can abstract tiny variations in the distance between eyes or the vertical positioning of mouth, noses, and eyes.

A complicated skill, like facial recognition, depends on having the ability to extract deeper, more abstract information from the flood of information that comes into our senses when we recognize a face. Instead of simply recording the light, sounds, and smells from any one scene, we must respond to deep, abstract properties of the world we are experiencing. This allows us to detect extraordinarily subtle and complex similarities and differences in situations that allow us to act effectively, even in new situations we never encountered before.

Abstract information is helpful because it can be used to guide us to pick out the information we are interested in from an incredibly complex array of possibilities. What is true of facial recognition is also true of other perceptual experiences. We make use of abstract information, for example, to recognize familiar melodies. Once you have heard Brahms' Lullaby, you can recognize it no matter what key it is played in or what instrument it's played on, even if it's played with several errors or mistakes. Whatever it is that allows us to recognize a familiar tune, it is not a memory of the specific experience of hearing that tune in the past. It must be something quite abstract. We rely on such abstract information in the act of recognition all the time, despite the fact that we are not even aware of it.

Borges understood that remembering everything is in conflict with what our minds do best — *abstract* information from the rich array of our experiences. This is why Funes described his mind as a "garbage heap." It is so filled with specific details that it is impossible to generalize or to comprehend. He cannot understand that all his encounters with four-legged, furry creatures are encounters with the same kind of animal:

> "He was, let us not forget, almost incapable of ideas of a general, Platonic sort. Not only was it difficult for him to comprehend that the general symbol *dog* embraces so many unlike individuals of diverse size and form; it bothered him that the dog from the time of 3:14 (seen from the side) should have the same name as the dog at the time of 3:15 (seen from the front)."

The reason most of us are not hyperthymesic is because it would make us less successful at what we were evolved to do. Our minds are busy trying to choose actions by picking out the most useful information from our experiences and leaving the rest behind. Remembering everything gets in the way of focusing on deeper abstract principles that allow us to recognize how new situations resemble past situations and what kinds of actions will be effective. Storing details is often unnecessary to act effectively; a broad picture is generally all we need. Sometimes storing details is counterproductive, as in the case of hyperthymesics and *Funes the Memorious*, whose memories were like garbage heaped with details. However, when solving problems, it is helpful for us to remember important information, and at times, our memories may not be where we would like them to be. For strategies about how to improve your memory, refer back to Chapter 5.

Logic of Action

If we had evolved in an environment that favored other capacities rather than to choose effective actions, we would probably have followed a different kind of logic than we do. If we evolved in a world that rewarded gambling on games of chance, we would probably be able to reason flawlessly about probability distributions and laws of statistics. If we had evolved in a world defined by deductive reasoning, we would probably be masters of reaching deductive conclusions, like Sherlock Holmes. Most of us, like Dr. Watson, are miserable at both of these mental activities (which take considerable training to master). Instead, we evolved in a world ruled by the *logic of action*. The logic of action involves imagining the consequences of our actions — what will happen when we rub a match across a rough surface, going out in the rain without an umbrella, or saying the wrong thing to a sensitive friend. In such situations, we imagine the world in some state and then, imagine the effects of an action that changes that state. There are other kinds of reasoning that we do not find easy or natural. It is hard to reason about $\sqrt[3]{8743}$ (i.e., the cube root of 8743); it is hard to reason about quantum mechanics; it's hard to predict the odds of winning the next time we place a bet at a casino. In fact, it can be difficult to engage in spatial reasoning: Is Reno, Nevada, east or west of Los Angles, California? Is Detroit, Michigan, north or south of Toronto, Canada? We are not good at everything and require specific strategies or a perspective to guide us with such difficult situations. We do excel at reasoning about how the world works — causal reasoning about effective action.[32]

For example, we easily see and can understand the simplest logical principle, *Modus Ponens.* See below for examples in the abstract and with content (also see Chapter 3):

Abstract	With Content
1. If A, then B.	1. If it rains, I will be wet.
2. A.	2. It rained.
3. Therefore B.	3. Therefore I am wet.

No problem; this is a valid way to think and the conclusion is correct. If one event occurs, a specific, second event will follow. This logic underlies our simplest associative thinking and dictates how we perceive and understand much of our experience in terms of action. Yet, this simple kind of thinking can lead us astray. Consider the following reasoning:

Abstract	With Content
1. If A, then B.	1. If it rains, I will be wet.
2. B.	2. I am wet.
3. Therefore A.	3. Therefore, it is raining.

Most people think this is a valid way of thinking and the conclusion is correct when presented with the abstract form of the problem, yet it is not. This way of thinking is referred to as the logical fallacy of *affirming the consequent.* That is, we affirm or accept that the consequent of "being wet" implies that the first premise, that "it is raining" is true. Being wet, however, can be the result of many other events: we may have fallen into a lake; we may have taken a shower, etc. Modus Ponens is obvious, easy to see, and to understand; one event (being wet) follows from a prior event (it rained). In this way of thinking, we are reasoning causally in a forward direction. Affirming the consequent

(I'm wet, therefore it must be raining) is not obvious and requires that we reason backwards from a state of "being wet" to what might be a possible cause of being wet, rain. Forward causal reasoning is obvious and easy to understand and seems intuitive; causal reasoning backwards, diagnostically, requires more than observing a sequence of events. It requires consideration of counter examples (what if it were not raining?) and alternative possibilities (what else could make you wet?). These two ways of thinking, causally forward and diagnostically back ways, illustrate two very different ways of thinking about experience. The first, thinking causally forward, is easy and seems quite natural. The second, thinking diagnostically, is hard and usually requires time, a strategy, or some training as a guide.

Causal Reasoning Forward and Backward

Causal reasoning is the basis of human cognition. It is what our minds do best. Yet, not all aspects of it are equally easy. We can reason both forward and backwards. Forward reasoning is thinking about how causes produce effects. We use it to predict the future, how events today will cause events tomorrow. We also use it to figure out how things work; how, for example, pushing certain keys gets a computer to perform specific functions. The example of the logical principle modus ponens requires forward reasoning. Reasoning backwards is reasoning from effect to cause. Doctors do it to diagnose the cause of symptoms and mechanics do it to diagnose what's wrong with your car. Backward causal reasoning involves figuring out how something that has happened came about. It's easier for us to reason forward — from cause to effect — than backwards from effect to cause. For instance, it is easier for a doctor to predict that someone with a peptic ulcer will have abdominal pains than it is to reach the conclusion that someone with abdominal pains has a peptic ulcer — there are many other possible causes of abdominal pains to consider. Backward causal reasoning, from effect to possible causes, requires more time and can be difficult. But, nonetheless, backward reasoning is what makes us special as humans; it is not clear that any other living organism can reason backwards as effectively as humans do.

The idea of reasoning backwards can be very helpful in solving mathematical problems. On the surface, the very title of this problem-solving strategy sounds confusing. This stems from a lack of familiarity with the procedure. From their earliest days in school, students are typically taught to solve problems in the most straightforward way possible. This is the way typical mathematics textbook problems are intended to be solved. Unfortunately, a substantial portion of this supposed "problem solving" is done by rote. Students struggle through one problem in a section, the teacher then usually reveals a "model solution," and the remaining "problems" in the section are solved in a similar manner. Little imaginative thinking is required of the students. In fact, we do not even consider these as *problems* — instead we refer to them as *exercises*, whose purpose is simply to reinforce a particular method of solution via repeated use. When, in the typical high school geometry course, students are first required to write proofs, they once again look for ways in which they can merely repeat previous procedures to solve successive problems. We seek to enable students to question the value of learning mathematics merely by rote. Such a student would then be more receptive to the problem-solving strategy offered here, namely, *working backwards.*

Developing a time schedule is another real-life example of using the working-backward strategy. When people develop a schedule for various tasks that must be completed by a certain time, they often start with what has to be done, the time at which all the work must be completed, and how long each task should take. Then, working backwards, they assign time "slots" to each task, and thus to arrive at the appropriate time to begin the work (see Chapter 5 for more information about how to go about doing this).

The working-backwards strategy is also widely used every day in traffic investigations. When the police are confronted by an automobile accident, they must begin to work backwards from the time of the accident to see what were the causes, which car swerved immediately before the collision, who hit whom, which driver was at fault, what were the weather conditions at the time of the accident, and so on, as they attempt to reconstruct the accident.

When we look at the procedures which students are shown in many of their typical textbook exercises, some very useful techniques are sometimes presented. Unfortunately, these are often taken for granted and not called to the students' attention. Students may be required to reason in the reverse order, even though they have not been told to do so. An obvious example, once again, is the procedure students should use when writing proofs in the high school geometry course. They should begin by examining what they are trying to prove before doing anything else. Thus, an attempt to prove line segments congruent might stem from proving a pair of triangles congruent. This, in turn, should suggest that the students look for the parts necessary to reach this triangle congruence. Continuing in this manner the student will be led to examine the given information. They are, in essence, *working backwards*. When the goal is unique, but there are many possible starting points, a clever problem solver begins to work backwards from the desired conclusion to a point where the given information is reached.

We must stress here that, when there is a unique endpoint (that which is to be proved) and a variety of paths to get to the starting point, the working-backward strategy may be desirable. The "working-forward" method is still the most natural method for solving the problem. In fact, working in the forward direction is used to solve *most* problems. We are not saying that all problems should be attempted by the working-backward strategy. Rather, after a natural approach (usually "forward") has been examined, a backward strategy might be tried to see if this provides a more efficient, more interesting or more satisfying solution to the problem.

The best approach to determine the most efficient route from one city to another depends upon whether the starting point or the destination (endpoint) has more access roads. When there are fewer roads leading from the starting point, the forward method is usually superior. However, when there are many roads leading from the starting point and only one or two from the destination, an efficient way of planning the trip is to locate this final destination on a map, and determine which roads lead most directly back towards the starting position. Progressively, continuing in this way (that is, working backwards), one reaches a familiar road that is easily reachable from the starting point. At this step, you will have mapped out the trip in a very systematic way.

To display a real-life situation in which working backward is beneficial, consider a salesman who has an appointment in a distant city and he must determine the flight he will take in order to arrive comfortably on time for his meeting; yet not too far in advance. He begins by examining the airlines schedule, starting with the arrival time closest to his appointment. Will he arrive in time? Is he "cutting it too close?" So, he examines the next earlier arrival time. Is this time all right? What if there is a weather delay? When is the next earlier flight? Thus, by working backwards, the executive can decide the most appropriate flight to take to get to his appointment on time.

The strategy game of "NIM" is another excellent example of when it is appropriate to use the working-backward strategy. In one version of the game, two players are faced with 32 toothpicks placed in a pile between them. Each player, in turn, takes 1, 2, or 3 toothpicks from the pile. The player who takes the final toothpick is the winner. Players develop a winning strategy by working backwards from 32 (i.e., to win, the player must pick up the 28th toothpick, the 24th toothpick, etc.) Proceeding in this manner from the final goal of #32, we find that a player can win if he or she picks the 28th, 24th, 20th, 16th, 12th, 8th, and 4th toothpicks. Thus, a winning strategy is to permit the opponent to go first, and proceed as we have shown.

Although many problems may require some reverse reasoning (even if only to a small extent), there are some problems whose solution is dramatically facilitated by working backwards. Consider the following problem; beware that it is not typical of the school curriculum, but rather a dramatic illustration of the power of working backwards.

Find a path on the adjoining grid beginning at "start" and ending at "end," where the sum of the cells is 50. You may pass through any open gate, after which the gate closes (see Figure 6.1).

Solution: Naturally, by a trial-and-error method, working forward, you should eventually stumble onto the right path. However, by using the

Start

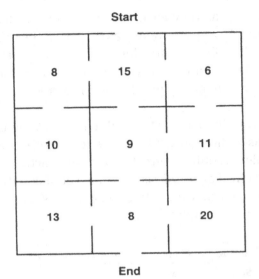

End

Figure 6.1

working-backward strategy you can simplify the problem significantly. You should be quick to realize that regardless of the path you take, you will have to pass through the 15 and 8 cells near the start and end, respectively. This means that $15 + 8 = 23$ will have to be used. This leaves us with a sum of $50 - 23 = 27$ that still needs to be visited. It should not be too difficult now to see that $8 + 10 + 9 = 27$, and thus, the required path is

$$\text{Start } 15 \rightarrow 8 \rightarrow 10 \rightarrow 9 \rightarrow 8 \text{ End}$$

It was the working-backward strategy that makes this problem far more manageable than by simply using trial and error and working forward from the start to the end.

The following is another problem that can be more easily solved by working backward.

Evelyn, Henry and Al play a certain game. The player who loses each round must give each of the other players as much money as the player has at that time. In round 1, Evelyn loses and gives Henry and

Al as much money as they each have. In round 2, Henry loses, and gives Evelyn and Al as much money as they each then have. Al loses in round 3, and gives Evelyn and Henry as much money as they each have. They decide to quit at this point, and discover that they each have $24. How much money did they each start with?

Solution: One usually begins this problem by setting up a system of three equations in three variables. Can it be done? Of course! However, since the problem requires a great deal of subtraction and simplification of parenthetical expressions, the final set of equations is likely to be incorrect. Even if they do obtain the correct set of equations, they must then be solved simultaneously.

Round	Evelyn	Henry	Al
Start	x	y	z
1	$x-y-z$	$2y$	$2z$
2	$2x-2y-2z$	$3y-x-z$	$4z$
3	$4x-4y-4z$	$6y-2x-2z$	$7z-x-y$

This leads us to the following system of equations:

$$4x - 4y - 4z = 24,$$
$$-2x + 6y - 2z = 24,$$
$$x - y + 7z = 24.$$

Solving the system leads to $x = 39$, $y = 21$, $z = 12$. Thus, Evelyn began with $39, Henry began with $21, and Al began with $12.

We should understand that the problem stated the situation at the end of the story ("They each have $24.") and asked for the starting situation ("How much money did they each start with?"). This is almost a sure sign that the *working-backward* strategy could be employed. Let's see how this makes our work easier. We begin at the end with each having $24.

	Evelyn	Henry	Al
End of Round 3	24	24	24
End of Round 2	12	12	48
End of Round 1	6	42	24
Start	39	21	12

Evelyn started with $39, Henry with $21, and Al with $12, the same answers as we arrived at by solving the problem algebraically.

Intuitive and Deliberative Thought

This distinction between two different kinds of thought can be found throughout classical and modern philosophy and psychology. Daniel Kahneman, the Nobel Laureate, has recently revived this distinction in his book *Thinking, Fast and Slow.*[33] This distinction has been referred to by a variety of names: associative versus rule based, system 1 versus system 2, intuitive and deliberate, as well as fast versus slow thinking. Consider an animal whose name begins with the letter *e.* "Elephant" leaps to mind for most everyone. Intuitive thinking provides an immediate answer. No effort necessary; we are barely conscious of how we came to this answer. Or, consider unscrambling the anagram, *initiutve*; the word "intuitive" magically appears, without being conscious of how we produced the response. Now, consider the more difficult anagram, *vaeertidebli.* If you solved it, you were conscious of the result and also of the process by which you got there. You could almost see the process in which you move letters around (to get "deliberative"). Similarly, when solving a difficult arithmetic problem, you are conscious of each step along the way. With regard to causal reasoning, intuitive thinking involves thinking in a forward direction (the letter *e* brings to mind animals whose names begin with that letter, elephant). While deliberative thinking typically involves backward causal reasoning (when trying to solve *vaeertidebli*, we may need to begin by thinking of starting the correct word with letters other than *vae*). Furthermore, the conclusions we come to quickly and intuitively are not always the same as the conclusions we come to through careful deliberation. Often the conclusions we come to intuitively are in conflict and may be overruled by conclusions that we come to with more slow and effortful deliberation. Intuition leads to one quick conclusion but deliberation makes us hesitate.

We can see this difference when we are asked if we know how everyday objects work. When people are asked how everyday things such as bicycles, zippers, and toilets work, they typically assert that they do. Yet when asked to provide a detailed explanation of how a bicycle, a zipper, or a toilet actually works, their explanations are superficial and often incorrect. This distinction is known as the *illusion of explanatory depth*.[34] This illusion has been easily documented with a very simple method. Participants are asked a series of simple questions:

1. On a scale from 1 to 7, how well do you understand how a zipper works?

2. How does a zipper work? Describe in as much detail as you can all the steps involved in a zipper's operation.

3. Now, on the same 1 to 7 scale, rate your knowledge of how a zipper works.

If you are like most of the people studied, you assert with confidence an understanding of zippers. However, if you don't work in a zipper factory, you would have little to say in response to the second question. Then when asked to respond to the third request to rate your understanding of a zipper, you would downgrade your rating. Respondents consistently show a little more humility by lowering their rating. After trying to explain how a zipper works, most people realize they have little idea and lower their rating of their knowledge by a point or two. By their own admission, respondents thought they understood how zippers work better than they did. When people lowered the ratings of their knowledge, they were essentially saying, "I know less than I thought I did." Similar results were found with undergraduate and graduate students from both elite and regional universities when asked about piano keys, flush toilets, cylinder locks, helicopters, quartz watches, and sewing machines. Likewise, this very robust finding occurs when asked about complex issues such as tax policies, foreign relations, GMO's, climate change, and even their own finances.

One interpretation of what occurs with the illusion of explanatory depth is that people overestimate how much they know to the first question about how things work because their first answer is based on intuitive understanding of what happens. Intuitive casual reasoning provides a superficial assessment of what people know about how things operate. Asking people to explain in detail how something works, forces them to think more deliberately about the operations and what causes them to work. After this shift in thinking, they re-evaluate their knowledge and lower their ratings of what they know. The illusion of explanatory depth is a product of the intuitive mind; we think about how things work automatically and effortlessly. But when we deliberate the illusion is shattered.

Another way to see the distinction between intuitive and deliberative thought is with the *Cognitive Reflection Test*. This test consists of three problems. The first is:

> A bat and a ball cost $1.10. The bat costs one dollar more than the ball. How much does the ball cost?

The answer 10 cents comes to mind almost immediately for most people. The real question is whether you just accept your first intuition about the answer or did you check it. If you check it, you'll see that if the ball costs 10 cents and the bat costs $1.00 more than the ball, then the bat costs $1.10, and together the cost comes to $1.20. So, the answer is not 10 cents. A small portion of people who do check their answers, realize that 10 cents is wrong and calculate the right answer (5 cents). These people suppress their intuitive response and deliberate before responding.

Another such situation can be found if you are going to shop at the store that advertises 10% discount on all its items year-round and on a special holiday has a discount on top of that of 20%. Most people intuitively believe that they are getting a 30% discount, when in fact, the two successive discounts of 10% and 20% are actually equivalent to a discount of 28%, since the 20% discount is coming off an already discounted price, and therefore, is less a discount than one would have if the 20% were taken off the original price.

The second problem provides a similar conflict between intuitive and deliberate thinking.

In a lake, there is a patch of lily pads. Every day the patch doubles in size. If it takes 48 days for the patch to cover the entire lake, how long would it take for the patch to cover half of the lake?

For most people the answer "24" comes to mind. If the pad doubles in size every day, then if the lake is half covered on day 24 it would be fully covered on day 25. But the problem states that the lake is fully covered on day 48. So, 24 cannot be the correct answer. The right answer must be one day before it is fully covered, day 47. This last problem is similar to the first two:

If it takes 5 machines 5 minutes to make 5 widgets, how long would it take 100 machines to make 100 widgets?

The answer is not 100, although most people think so. If each machine takes 5 minutes to make one widget, then the correct answer is 5 minutes. Each machine making one widget in 5 minutes gets you 100 widgets.

What these three problems have in common is that an incorrect answer immediately pops to mind. To get the right answer, the intuitive answer must be suppressed, and you must do a little calculation. Most people don't bother. Rather than suppressing the intuitive answer and engaging in the little bit of deliberation to figure out the right answer, people just blurt out the intuitive answer, the first answer that comes to mind. Less than 20% of a very large United States sample gets the three problems right. Mathematicians and engineers do better than poets and painters; but not much better. About 40% of students at the Massachusetts Institute of Technology got all three problems correct, and only 26% of Princeton students did so.

The Cognitive Reflection Test distinguishes people who reflect before they answer from those who answer with the first thing that come to mind. People who reflect depend on more deliberative thinking; those who are less reflective depend more on intuitions. People who are more reflective tend to be more careful when given problems that involve reasoning, make fewer errors, and are less likely to fall for tricks than those people that are more intuitive. They are less impulsive, take more risks proposing solutions when working on a problem, and will wait longer for a greater reward. Other studies found that reflective people crave details. When shown a variety of advertisements for products that differed in the amount of detailed description for each product, participants who were more reflective preferred products with more detail in the advertisement. Those scoring lower on the Cognitive Reflection Test preferred descriptions with much less detail. And, most relevant to this discussion, more reflective people (who score higher on the test) show less of an illusion of explanatory depth than less reflective people. Intuition gives us a simplified, coarse analysis, creating the illusion that we know more than a fair amount. But when we deliberate, we come to appreciate how complex things are, revealing how little we actually know.

Neglecting Base Rates and Covariation

At the heart of intuitive thinking is covariation: x and y covary, if x tends to be present whenever y is; and if x is absent y is absent. Covariation is important in making predictions and is necessary in establishing cause and effect (e.g., see Chapter 3). Education leads to higher paying jobs; a good breakfast makes you feel better throughout the day. While this way of thinking is intuitive and feels natural, it can lead to errors and invalid conclusions. We can sometimes think there are relationships where there are only questionable ones or none at all. Are you more likely to fall in love with someone who is tall? Do wearing vertical stripes make you look thinner? Does your car start more easily when you pump the gas pedal? Does the elevator come more quickly if you continuously keep pushing the button? When we see covariation when there is none, we are tricked by the *illusion of covariation,* which is quite common.

The illusion of covariation was demonstrated in a classic study using the Rorschach test.[35] In the Rorschach test, people are shown inkblots and asked to describe them. Clinical psychologists then examine the descriptions looking for patterns among different types of response. For instance, mentioning motion in a description of a Rorschach card indicates imagination and a rich inner life; responses that describe the white spaces in the cards indicate rebelliousness. In this study, fictitious responses were paired with descriptions of the people who supposedly made the responses. For example, one transcript was attributed to a man who believes others are out to get him; another was attributed to a man who has "sexual feelings towards other men." After randomly pairing the transcripts with the personality descriptions, the pairs were shown to undergraduates, who had no experience with the Rorschach test, and a group of trained clinical psychologists, who had extensive experience with the Rorschach test. The surprise finding was that both groups saw covariation between certain responses and certain personality descriptions despite the fact that there were no such relationships among the information that was examined. For example, both groups saw that certain responses were indicative of sexual orientation. For both groups this pattern was illusory, not present in the actual pairings. The two groups did not differ in seeing the likelihood of the pattern that really wasn't there. Both inexperienced undergraduates and experienced clinicians were caught by the same illusion of covariation. Similar findings of illusory covariation occur among financial planners making claims about investments and among physicians diagnosing cancer. These errors occur even when people are doing their best to be careful (when offered cash bonuses for performing accurately).

The explanation for the persistence of the illusion of covariation is that people seem to consider only a subset of all the evidence that is presented, a subset that is biased by their prior expectations. This biased evaluation of the evidence virtually guarantees mistaken judgments. As mentioned earlier, this selective process is referred to as the *confirmation bias*; a tendency to be more responsive to evidence that confirms your beliefs, rather than evidence that might challenge your beliefs (see Chapter 3). For example, if you believe that big dogs are vicious, you're more likely to notice big dogs that are, in fact, vicious and little dogs that are friendly. When such a sample of dogs is available, these are the dogs you perceive and remember, and fail to notice big dogs that are friendly and small dogs that are vicious. When asked to estimate covariation between dog size and temperament, you'll overestimate the relationship.

Assessment of covariation can be understood as a result of neglecting *base-rate information* about how frequently something occurs in general.[36] Imagine testing a new drug to cure a certain disease, and let's say a study shows that 70% of the patients taking the drug recover from the disease. By itself this information is uninterruptable, because we need to know in general how often people recover from this disease. If it turns out that the overall rate of recovery is 70%, without taking the drug, the new drug is having no effect at all. In a classic study, it was found that people prefer easy, obvious answers over base-rate information.[37] Participants were asked:

> If someone is chosen at random from a group of 70 lawyers and 30 engineers, what is his profession likely to be?

Another group of participants were asked the same question but without the base-rate information and, instead, were given a brief character description of the chosen individual. Some of the descriptions were stereotypes to suggest that the person is a lawyer or an engineer, and some were neutral.

Not surprisingly, people used the type of information that was available. In the first group they used the base-rate information; in the second group they used the character descriptions to help estimate the probability of a certain profession. In a third group, who were given both kinds of information, you'd expect that participants would use both base-rates and the thumbnail sketches for their estimations. But instead this third group relied only on the descriptive information and ignored the base-rates. They responded the same way, based on the descriptions, regardless if the base-rate was 70 lawyers and

30 engineers or 30 lawyers and 70 engineers. Thus, when asked if Tom is a lawyer, participants asked themselves how much Tom resembles their idea of a lawyer and ignored the base-rate information. Making a judgment on resemblance is intuitive, quick, and effortless; making a judgment on base-rate information is deliberate requiring time and effort.

Conclusion

In sum, we can think reasonably about problems in two ways. Sometimes we rely on intuitive thinking that is quick and effortless, based on obvious information. While this type of thinking is often useful, it is prone to error. On other occasions, we think about problems in a more deliberate way. This type of thinking is slower, more effortful, and more accurate. When do we use intuitive thinking and when do we use the more accurate deliberative thought? To some extent, it may be a matter of choice to either accept intuitive thinking or shift to more deliberate thought.

Although deliberative thought is very powerful, when given a choice, we tend to opt for the easier and quicker way of thinking. The type of thinking we use may be triggered by certain cues. Time pressure when solving problems and making judgments can affect our tendency to utilize fast intuitive thought. In contrast, focusing on the process of judgment can initiate more deliberate thinking. For instance, experience shows that in a university course in Cognition, most students are instantly suspicious when the instructor asks a question that seems to have an easy, obvious answer. Factors like time pressure and focus of attention, however, can't be the whole story, because sometimes people who are rushed and distracted get things right and people who have time to work on a problem and are focused make errors. There are other cues that can trigger deliberate thought. For example, to counter the tendency to neglect base-rate information, this error can be avoided. We can heighten sensitivity to base-rates, if base-rate information is presented right away, when the problem is introduced. Also, we are more aware of base-rate information, if it is cast in terms of frequencies (12 out of a 1000) rather than when cast in terms of percentages (1.2%) or in decimal form (0.012). If chance is emphasized when the problem is introduced, we are more likely to realize that evidence maybe a fluke or accidental and not indicative of a reliable pattern. Moreover, judgments are more deliberate and accurate and less prone to intuitive error when evidence is presented in easily understood statistical terms. For example, it is relatively clear that an athlete's performance in the first quarter of a game is just a sample of evidence (and may not reflect her performance in other quarters). Also, performance in sporting events can be measured in terms of points. In contrast, it is harder for an employer to see that a 10-minute interview is just a "sample" of evidence, and that other impressions might come from other samples (like being interviewed on other days or seeing the person in another setting). And, it is difficult to quantify the employer's impressions during the interview. Thus, problem solving can be more or less accurate depending on how the problem is presented.

It is clear that the use of deliberate thinking when solving problems and making judgments depends on factors in the problem situation and in how the problem is presented. Using deliberate reasoning when problem solving requires deliberate education. We can be educated to use the skills necessary to see the complexity of a problem when working towards an accurate solution. Errors in judgment will always occur. But such errors are not the result of deep flaws in our capacity for making judgments. Instead, the errors arise largely because the problem situation does not trigger our deliberate thinking abilities. And, most importantly education can improve this state of affairs, allowing us to use better judgment in a range of settings. Education doesn't eliminate judgment errors, nor does it guarantee success of deliberative thought. But the right sort of education can decrease the danger of making such errors.

Endnotes

1. Fibonacci used the term "Indian figures" for the Hindu numerals, since that was where they originated.
2. For more on the Fibonacci numbers, see *The Fabulous Fibonacci Numbers*, by A. S. Posamentier and I. Lehmann, Amherst, NY: Prometheus Books, 2007.
3. Danesi, M. (2002). *The Puzzle Instinct: The Meaning of Puzzles in Human Life*. Bloomington, IN: Indiana University Press.
4. Luchins, A. S. & Luchins, E. H. (1959). *Rigidity of Behavior: A Variational Approach to Einstellung*. Eugene, OR: University of Oregon Press.
5. Bartlett, F. (1958). *Thinking: An Experimental and Social Study*. New York: Basic Books.
6. Duncker, K. (1977). On problem solving. In P. C. Wason and P. N. Johnson-Laird (eds.), *Thinking: Reading in Cognitive Science*. Cambridge: Cambridge University Press.
7. Kohler, W. (1925). *The Mentality of Apes*. New York: Harcourt Brace Jovanovich.
8. Wertheimer, M. (1959). *Productive Thinking*. New York: Harper & Rowe.
9. Polya, G. (1957). *How to Solve It?* Princeton, NJ: Princeton University Press.
10. Newell, A. & Simon, H. (1972). *Human Problem Solving*. Englewood Cliffs, NJ: Prentice-Hall.
11. Reisberg, D. (2013). *Cognition: Exploring the Science of the Mind*. New York: W. W. Norton.
12. Bassock, M. & Holyoak, K. (1989). Interdomain transfer between isomorphic topics in algebra and physics. *Journal of experimental Psychology: Learning, Memory, and Cognition*, 15, 153–166.

13. You can play a card version of the Tower of Hanoi game. Take four cards of the same suit (spades), in numerical order — ace, 2, 3, and 4. Put the cards in a space, calling it A. Set up two empty spaces, B and C, right next to the cards. The object is to relocate the cards to space C in accordance with the same rules: a larger value card may never be placed on top of a smaller valued card; only one card at a time can be moved to a new space.

14. Meyer, R. E. (1992). *Thinking Problem Solving, Cognition*. New York: Worth Publishing.

15. Polya, G. (1945). *How to Solve It?* Princeton, NJ: Princeton University Press.

16. Wertheimer, M. (1945/1982). *Productive Thinking*. Chicago, IL: University of Chicago Press, for one account of this legend.

17. Safren, M. A. (1962). Associations, set, and the solution to word problems. *Journal of Experimental Psychology*, 64, 40–45.

18. Levine, M. (1994). *Effective Problem Solving*. Upper Saddle River, NJ: Prentice-Hall.

19. Akin, E. (1996). Logic of errors. In A. S. Posamentier and W. Schultz (eds.), *The Art of Problem Solving: A Resource for Mathematics Teachers*. Thousand Oaks, CA: Corwin Press.

20. Hartman, H. (1996). Cooperative approaches to mathematical problem solving. In A. S. Posamentier and W. Schultz (eds.), *The Art of Problem Solving: A Resource for Mathematics Teachers*. Thousand Oaks, CA: Corwin Press.

21. Tversky, A. & Kahaneman, D. (1973). Availability: A heuristic for judging frequency and probability. *Cognitive Psychology*, 5, 207–232.

22. Schwarz, N., Bless, H., Strack, F., Klump, G., Ritternauer-Schatka, H. & Simon, A. (1991). Ease of retrieval as information: Another look at the availability heuristic. *Journal of Personality & Social Psychology*, 61, 195–202.

23. Tversky, A. & Kahaneman, D. (1974). Judgment under uncertainty: Heuristics and biases. *Science*, 185, 1124–1131.

24. Nisbitt, R. & Ross, L. (1980). *Human Inference: The Strategies and Shortcomings of Social Judgment*. Englewood Cliffs, NJ: Prentice-Hall.

25. Wason, P. (1968). Reasoning about a rule. *Quarterly Journal of Experimental Psychology*, 20, 271–281.

26. Wason, P. & Johnson-Laird, P. (1972). *Psychology of Reasoning: Structure and Content*. Cambridge, MA: Harvard University Press.

27. Griggs, R. & Cox, J. R. (1982). The elusive thematic materials effect in Wason's selection task. *British Journal of Psychology*, 73, 407–420.

28. von Neuman, & Morgenstern, O. (1947). *The Theory of Games and Economic Behavior*. Princeton, NJ: Princeton University Press.

29. Tversky, A. & Kahneman, D. (1987). Rational choice and the framing of decisions. In R. Hogarth and M. Reder (eds.), *Rational Choice: The Contrast Between Economics and Psychology*. Chicago, IL: University of Chicago Press.

30. Luis Borges, J. (1998). *Collected Fictions* (trans. Andrew Hurley). New York: Viking Press. Also see *Borges and Memory: Encounters with the Human Brain* (2012). Rodrigo Quian Quiroga. Cambridge MA: MIT Press.

31. Maurer, D. & Mondloch, C. J. (2002). The many faces of configural processing. *Trends in Cognitive Science*, 6(6), 225–260.

32. Cummins, D. D., Lubart, T., Alksins, O. & Rist, R. (1991). Conditional reasoning and causation, *Memory and Cognition*, 19(3), 274–282.

33. Kaheman, D. (2011). *Thinking Fast and Slow*. New York: Farrar, Straus.

34. Rozenblit, L. & Keil, F. (2002). The misunderstood of folk science: An illustration of explanatory depth. *Cognitive Science*, 26(5), 521–562. Also, see Keil, F. & Wilson, R. A. (2000). *Explanation and Cognition*. Cambridge, MA: MIT Press.

35. Chapman, J. & Chapman, L. J. (1971). Test results are what you think they are. *Psychology Today*, 5, 106–110.

36. Fredricks, S. (2005). Cognitive reflection and decision making. *Journal of Economic Perspectives*, 19(4), 25–42.

37. Kaheman, D. & Treversky, A (1973). On the psychology of prediction. *Psychological Review*, 80, 237–251.

Index